AA002328

2019 IEEE Workshop on Microelectronics and Electron Devices (WMED 2019)

Boise, Idaho, USA
26 April 2019

IEEE Catalog Number: CFP19564-POD
ISBN: 978-1-5386-1202-6

**Copyright © 2019 by the Institute of Electrical and Electronics Engineers, Inc.
All Rights Reserved**

Copyright and Reprint Permissions: Abstracting is permitted with credit to the source. Libraries are permitted to photocopy beyond the limit of U.S. copyright law for private use of patrons those articles in this volume that carry a code at the bottom of the first page, provided the per-copy fee indicated in the code is paid through Copyright Clearance Center, 222 Rosewood Drive, Danvers, MA 01923.

For other copying, reprint or republication permission, write to IEEE Copyrights Manager, IEEE Service Center, 445 Hoes Lane, Piscataway, NJ 08854. All rights reserved.

*** This is a print representation of what appears in the IEEE Digital Library. Some format issues inherent in the e-media version may also appear in this print version.

IEEE Catalog Number:	CFP19564-POD
ISBN (Print-On-Demand):	978-1-5386-1202-6
ISBN (Online):	978-1-5386-0571-4
ISSN:	1947-3834

Additional Copies of This Publication Are Available From:

Curran Associates, Inc
57 Morehouse Lane
Red Hook, NY 12571 USA
Phone: (845) 758-0400
Fax: (845) 758-2633
E-mail: curran@proceedings.com
Web: www.proceedings.com

2019 IEEE Workshop on Microelectronics and Electron Devices

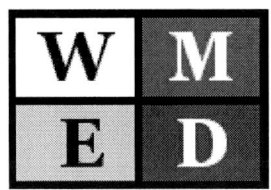

Jordan and Simplot Ballrooms
Student Union Building

Boise State University
Boise, ID, USA

April 26th 2019

This workshop is receiving technical co-sponsorship support from the IEEE Electron Devices Society.

2019 IEEE Workshop on Microelectronics and Electron Devices (WMED)

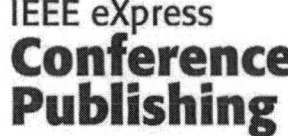

Produced by **IEEE eXpress Conference Publishing**

For information on producing a conference proceedings and receiving an estimate, contact conferencepublishing@ieee.org
http://www.ieee.org/conferencepublishing

Welcome to the IEEE WMED 2019

Dear Participant,

On behalf of the Organizing Committee, it is my pleasure to welcome you to the seventeenth annual Workshop on Microelectronics and Electron Devices (WMED 2019); a professional workshop hosted by the Boise Chapter of the IEEE Electron Devices Society (EDS). WMED has long been recognized as one of the premiere technical workshops bringing world renowned researchers to the Mountain West to discuss the progress of semiconductor related technologies. We have an excellent technical program this year including invited talks and contributed papers and posters. Topics include "DRAM Scaling Challenges and Future", "CDM ESD Protection", "Emerging Memory Technologies for Neuromorphic Computing", "Phase-Change for Storage Class Memory: Novel Bonds at Work", "Bridging Mechanics and Electrochemistry in Battery Materials", "Nanoelectronics and Heterogeneous Integration with 2D Materials", "Innovations in Device Packaging: Glass, Lasers, Materials, and Integration", "Emerging Materials for Additive Manufacturing of Flexible Hybrid Sensors", and "Energy Efficient Circuits". I'd like to express my sincere appreciation to each of our speakers for their contributions to our technical program.

During the workshop, we provide a unique program for local high school students. The WMED High School Program introduces students to STEM career choices and features invited talks, panel discussions, and hands-on activities related to engineering. This year's focus for the program is Virtual Reality (VR). I'd like to specifically acknowledge Laurie Anderson of the Micron Foundation for her continued support and contributions to this program.

In addition, I would like to recognize and thank each of our Executive, Platinum, and Gold level sponsors for their generous financial support. Continued growth and success of WMED would not be possible without their contributions.

Finally, I wish to express my sincere thanks to each member of the 2019 organizing committee. The team has provided a valuable experience for attendees and continuing to improve WMED for the future. It's been my pleasure to work with such an excellent team.

I hope that you have a rewarding and enjoyable experience today. Please consider how you can contribute as well to WMED 2019, and we hope to see you next year.

Best Regards,
Ahmed Abdelnaby
General Chair, IEEE WMED 2019

IEEE WMED 2019 Organizing Committee

General Chair	*Ahmed Abdelnaby*
Technical Chair	*Adam Saxler*
Technical Vice Chair	*Durga Panda*
Program Chair	*Nancy Lomeli*
Publicity Chair	*John CaKa*
Finance Chair	*Yantao Ma*
Publications Chair	*Santanu Sarker*
Publications Vice Chair	*Chandra Tiwari*
Registration Chair	*Anand Ursekar*
Registration Vice Chair	*Xiao Li*
Sponsorship Chair	*Francois Fabreguette*
Sponsorship Vice Chair	*Robert Grubbs*
University Relations Chair	*Sepideh Rastegar*
High School Program Chair	*Becky Munoz*
High School Program Vice Chair	*Josiah Jebaraj*
Intern Relationship Chair	*Ashlie Petz*
TLP Coodinator	*Emily Berriochoa*
Webmaster	*Pratap Murali*
Technical Committee	Lan Li, Randy Wolf, Tim Hollis, Jaydeb Goswami, Ankit Sharma

Reviewers:

Andrea Ghetti, Andrea Redaelli, Arun Dhayalan, Daniel Billingsley, Deepak Chandra Pandey, Durga Panda, Enrico Varesi, Jian Li , Josiah Jebaraj Johnley Muthuraj, Kaushik Varma Sagi, Kehao Zhang, Pavan Anirudh Poosarla, Prathmesh Kadam, Raj Kumar Manthena, Santanu Sarkar, Sriram Balasubramanian, Sumeet Pandey, Suren Eruvuru, Zhao Zhao

IEEE WMED 2019 Technical Program
Friday, April 26[h], 2019
8:00AM - 5:00PM

Time		
7:30 AM	**WMED 2019 - Check In and Door Registration** With Continental Breakfast	
8:00 AM	**Welcome to Boise State University: Simplot Ballroom**	
8:05 AM	**Welcome to WMED 2019: Simplot Ballroom**	
8:10 AM	**Keynote Address: Simplot Ballroom** DRAM Scaling Challenges and Future *Shigeru Shiratake, Micron*	
9:10 AM	**Break**	
9:20 AM	**Invited Talk: Simplot Ballroom** Does CDM ESD Protection Really Work? *Albert Wang, University of California - Riverside*	**Invited Talk: Jordan AB** Phase-Change for Storage Class Memory: Novel Bonds at Work *Matthias Wuttig, RWTH Aachen*
10:05 AM	**Invited Talk: Simplot Ballroom** Emerging Memory Technologies for Neuromorphic Computing *Dmitri Strukov, UCSB*	**Invited Talk: Jordan AB** Bridging Mechanics and Electrochemistry: Experiments and Modeling on Battery Materials *Kejie Zhao, Purdue University*
10:50 AM	**Break**	
11:00 AM	**Keynote Talk: Simplot Ballroom** Nanoelectronics and Heterogeneous Integration with 2D Materials *Eric Pop, Stanford University*	
12:00 PM	**Buffet Luncheon: Jordan D**	
12:45 PM	**Poster Session: Jordan D**	
	Invited Talks	
1:30 PM	**Simplot Ballroom** Innovations in Device Packaging: Glass, Lasers, Materials, and Integration *Steve Groothuis, Samtec Microelectronics*	
2:15 PM	**Simplot Ballroom** Emerging Materials for Additive Manufacturing of Flexible Hybrid Sensors *David Estrada, Boise State University*	**Jordan AB** Phase Transition Materials Assisted Energy Efficient CMOS Design *Jaydeep Kulkarni, University of Texas - Austin*
3:00 PM	**Break**	
	Contributed Papers: (Parallel Sessions)	
3:20 PM	**Session 1: Simplot Ballroom**	**Session 2: Jordan AB**
3:20 PM	Numerical Study of Non-Circular Pillar Effect in 3D-NAND Flash Memory Cells *Albert Fayrushin, Micron*	Deep learning for industrial IoT-empowered processes: Methods, applications, infrastructure, and practical considerations *Sujata Butte, University of Idaho*
3:45 PM	Metal-Site Dopants in Two-Dimensional Transition Metal Dichalcogenides *Lan Li, Boise State University*	Resistance-Based Embedded Non-Volatile Memories as Synaptic Devices in Spiking Neural Networks *Foroozan Koushan, Micron*
4:10 PM	Structural and Magnetic Properties of CoPd Alloys for Non-Volatile Memory Applications *Joseph Abugri, University of Alabama*	Three Dimensional Time Domain Simulation of the Quantum Magnetic Susceptibility *Jennifer Houle, University of Idaho*
4:35 PM	3D Simulation Technique To Predict Failure of Photo Marks Interaction *Radhakrishna Kotti, Micron*	An optically reconfigurable gate array workable under a strong gamma radiation environment *Minoru Watanabe, Shizuoka University*
5:00 PM	**WMED 2019 End**	

IEEE WMED 2019 Table of Contents

Welcome to the IEEE WMED 2019 v

Organizing Committee vi

Technical Program vii

High School Program x

Keynote Address

DRAM Scaling Challenges and Future xiii

Invited Talks

Does CDM ESD Protection Really Work? xiv
Albert Wang

Neurocomputing and Hardware Security with Memory Devices xv
Dmitri B. Strukov

Phase-Change Materials for Storage Class Memory: Novel Bonds at Work xvii
M. Wuttig

Bridging Mechanics and Electrochemistry: Experiments and Modeling on Battery Materials xviii
Kejie Zhao

Keynote Talk

Nanoelectronics and Heterogeneous Integration with 2D Materials xix
Eric Pop

Invited Talks

Innovations in Device Packaging: Glass, Lasers, Materials, and Integration xx
Steve Groothuis

Emerging Materials for Additive Manufacturing of Flexible Hybrid Sensors xxi
David Estrada

Phase Transition Materials Assisted Energy Efficient CMOS Design xxii
Jaydeep Kulkarni

Invited Contribution

Does CDM ESD Protection Really Work? xxv
Mengfu Di, Han Wang, Feilong Zhang, Cheng Li, Zijin Pan, and Albert Wang

Contributed Paper Sessions xxix

Numerical Study of Non-Circular Pillar Effect in 3D-NAND Flash Memory Cells 1
Albert Fayrushin, Haitao Liu, Aurelio Mauri, Gianpietro Carnevale, Hyejin Cho, and Duo Mao

Metal-Site Dopants in Two-Dimensional Transition Metal Dichalcogenides 5
I. Williamson, M. Lawson, S. Li, Y. Chen, and L. Li

Structural and Magnetic Properties of CoPd Alloys for Non-Volatile Memory Applications 10
Joseph B. Abugri, Billy D. Clark, Sujan Budhathoki, Pieter B. Visscher, Adam Hauser, and Subhadra Gupta

3D Simulation Technique to Predict Failure of Photo Marks Interaction 15
Radhakrishna Kotti

Deep Learning for Industrial IoT-Empowered Processes: Methods, Applications, Infrastructure, and Practical Considerations .... N/A
Sujata Butte, Aleksandar Vakanski, and Min Xian

Resistance-Based Embedded Non-Volatile Memories as Synaptic Devices in Spiking Neural Networks 23
Foroozan S. Koushan and Sung-Mo Steve Kang

Three Dimensional Time Domain Simulation of the Quantum Magnetic Susceptibility 27
Jennifer Houle, Dennis Sullivan, Ethan Crowell, Sean Mossman, and Mark G. Kuzyk

An Optically Reconfigurable Gate Array Workable Under a Strong Gamma Radiation Environment 32
Shinya Fujisaki, Takumi Fujimori, and Minoru Watanabe

Poster Session 37

Atomic Layer Deposition of Gallium Phosphide 39
Sara Kuraitis and Elton Graugnard

Investigation of Molybdenum Disulfide Atomic Layer Deposition on Dielectric Surfaces 40
Steven Letourneau, Jake Soares, Anil U. Mane, Jeffrey W. Elam, and Elton Graugnard

Dependency of RGB and Metrology Data Correlation on Wafer Alignment 41
Swetha Barkam, James Cultra, Zhenyu Bo, and Sri Sai Vegunta

Scanning Probe Microscopy for Nanoscale Characterization of Electrical and Magnetic Properties 41
Olivia Maryon

Author Index 42

Sponsors 44

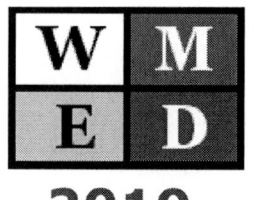

Workshop on Microelectronics and Electron Devices
High School Program
Friday, April 26, 2019
8:00 am – 2:00 pm
@ Boise State University

2019

2019 WMED Schedule

7:30AM **Simplot Ballroom Lobby**	**Check-in and Door Registration** **@ The High School Table** *With Continental Breakfast* Check in between 7:30 and 8:00 to get your nametag and breakfast.
8:00AM **Simplot Ballroom**	**Welcome to WMED 2019**
8:15AM **Simplot Ballroom**	**Keynote Address** **Memory: DRAM Scaling Challenges and Future** *Shigeru Shiratake, CVP, Micron Technology*
9:00 AM **Lookout Room**	**Welcome to WMED 2019 Student Session** *Laurie Anderson, K12 STEM Outreach Manager, Micron Foundation* *Becky Muñoz, Sr. YE Engineer, Micron* *Josiah Muthuraj, Sr. CMP Engineer, Micron*
9:15 AM **Lookout Room**	**Industry Demonstrations** **Virtual Reality, Augmented Reality and the Software that makes it happen** *Idaho Virtual Reality Council, Silverdraft, Blocksmith, Blackbox VR; VR1 Arcade*
10:15 AM **Lookout Room**	**Invited Talk:** *Ryan DeLuca, CEO and Co-Founder, Black Box VR*
11:30 AM **Lookout Room**	**Lunch** Pizza, salad and beverages will be provided. Vegetarian options available.
12:00 PM **Lookout Room**	**Invited Talk:** *Linda Somerville, CVP, Micron Technology*
1:00 PM **Lookout Room**	**Q&A with Engineering Panel** Open Q/A with Engineers from Industry, and/or University Engineering Professors, and/or College Engineering students

Keynote Address

Keynote Address

DRAM Scaling Challenges and Future

Shigeru Shiratake

Corporate Vice President, DRAM & EM Device Technology and Process Integration
Technology Development, Micron Technology, Inc., Boise, ID, USA

Abstract:

Over the last 10 years, memory and storage performance requirements have been driven by exponential growth in computing and data needs. The future brings continued demands for performance improvements as artificial intelligence, high performance computing and data analysis continue to become integrated into every aspect of modern life. Artificial Intelligence (AI) has become a game changer -- harvesting enormous data sets and converting them to insights and intelligent action in automotive, mobile edge device, enterprise storage, and medical imaging applications. Historically, DRAM memory technology has scaled and sustained the scaling factor to achieve enough cost per bit reduction and meet high-speed with low-power requirements. And, DRAM technology in High Volume Manufacturing (HVM) has been transformed with 10 nanometers class process technology. However, as future performance demands continue to increase, DRAM will run into many challenges achieving cell device below 1xnm technology. It will take innovative processes and material development to overcome these challenges and provide the breakthroughs and innovative design approaches to deliver on future performance demands and scaling requirements.

Speaker's Biography

Shigeru Shiratake is the Corporate Vice President for Technology Development's DRAM and Emerging Memory research development organization. He joined Micron in 2013 as the Section Director of the DRAM process integration organization. Prior to joining Micron, He was an executive of Elpida Memory (acquired by Micron in 2013) between 2005 – 2013, leading several DRAM programs. His experience spans 30 years in the memory technology sectors, and includes additional leadership positions with Renesas Technology Inc., and Mitsubishi Electric, Corp.

Shiratake is a published author of many technical papers, holds dozens of patents pertaining to semiconductor technology. He holds a Bachelor's and Master's of Science in Electronic Engineering from Yamaguchi University, Japan.

Invited Talk

Does CDM ESD Protection Really Work?

Albert Wang

Professor, Dept. of Electrical and Computer Engineering, University of California, 417 Winston Chung Hall, Riverside, CA 92521, USA. Email: aw@ece.ucr.edu

Abstract

On-chip electrostatic discharge (ESD) protection is required to protect ICs against ESD failures of various origins. While ESD protection against human body model (HBM) ESD failures has been very successful, charged device model (CDM) ESD protection design emerges as a new challenge for ICs at advanced technology nodes. This paper reports a comparison study of CDM ESD protection versus HBM ESD protection. It found that the current practices of using conventional pad-based ESD protection networks for CDM ESD protection may not work. This study leads to a game-changing question to the field: Are existing CDM ESD protection approaches fundamentally faulty? It therefore calls for a completely new on-chip CDM ESD protection method to truly resolve the real-world CDM ESD failure problems.

Speaker's Biography

Albert Wang received his BS degree from Tsinghua University and PhD degree in from the State University of New York at Buffalo, both in Electrical Engineering, in 1985 and 1996, respectively. Currently, he is a Professor of Electrical and Computer Engineering at University of California, Riverside. His research covers AMS/RF ICs, integrated design-for-reliability, 3D heterogeneous integration, emerging nano devices and circuits, and LED visible light communications. He published one book and 260+ peer-reviewed papers and holds fifteen U.S. patents. His editorial board services include IEEE *Transactions on Circuits and Systems I*, IEEE *Electron Device Letters*, IEEE *Transactions on Circuits and Systems II*, IEEE *Transactions on Electron Devices*, *IEEE Journal of Solid-State Circuits* and *Journal of Engineering*. He has been IEEE *Distinguished Lecturer* for IEEE Electron Devices Society, IEEE Circuits and Systems Society and IEEE Solid-State Circuits Society. He is Sr. Past President (2018-2019) and was President (2014-2015) for IEEE Electron Devices Society. He was Chair for the *IEEE CAS Analog Signal Processing Technical Committee (ASPTC)* and member of the *International Technology Roadmap for Semiconductor (ITRS)* Committee. He is a member for *IEEE 5G Initiatives Committee* and *IEEE Fellow Committee*. He was General Chair (2016) of *IEEE RFIC Symposium*. Wang received CAREER Award from National Science Foundation. Wang is IEEE Fellow, AAAS Fellow, and Fellow of National Academy of Inventors.

Invited Talk

Neurocomputing and Hardware Security with Memory Devices

Dmitri B. Strukov

UC Santa Barbara, Santa Barbara, CA 93106-9560, U.S.A.

Abstract

Recent advances in dense, continuous-state nonvolatile memories have enabled extremely fast, compact, and energy-efficient analog and mixed-signal circuits [1-8]. Such circuits are perfectly suited for hardware implementations of inference [9-13] and other mode advanced functionalities [10, 14-16] in neuromorphic networks, which require massive amounts of moderate precision dot-product operations. In the first part of my talk, I will review typical implementations of such mixed-signal circuits, and then describe some recent experimental demonstrations by my group of prototype mixed-signal neuromorphic networks based on metal-oxide ReRAMs [10,11,14,16] and floating gate memories [5,12,13].

The device to device variations and nonlinear I-V characteristics of nonvolatile memories also enable very efficient implementations of physical unclonable function, a key primitive in hardware security. In the second part of my talk, I will review our theoretical and experimental efforts toward developing such circuits [17-20].

Speaker's Biography

Dmitri Strukov is a Professor of Electrical and Computer Engineering at the University of California at Santa Barbara. Prior to joining UCSB Dmitri worked as a postdoctoral associate, first at Stony Brook University (Aug. 2006 – Dec. 2007), and then at Hewlett Packard Laboratories (Jan. 2007 – Jun. 2009) on various aspects of nanoelectronic devices and systems. He received a MS in applied physics and mathematics from the Moscow Institute of Physics and Technology in 1999 and a PhD in electrical engineering from Stony Brook University in New York in 2006. He is a member of ACM and IEEE societies.

Dmitri's research broadly concerns different aspects of computation, in particular addressing questions on how to efficiently perform computation on various levels of abstraction. The research spans across different disciplines including material science, device physics, circuit design, high level computer architecture, and algorithms with the emphasis on the emerging device technologies. Over the past decade, his major focus was on neuromorphic computing and, more recently, on hardware security implementations with resistive switching devices ("memristors") and floating gate memories.

References

[1] F. Merrikh Bayat et al., "Memory technologies for neural networks", in: Proc. IMW'15, Monterey, CA, May 2015, pp. 1-4.

[2] L. Ceze et al., "Nanoelectronic neurocomputing: Status and prospects", in: Proc. DRC'16, Newark, DE, Jun. 2016, pp. 1-2.

[3] G.C. Adam et al., "3-D memristor crossbars for analog and neuromorphic computing applications", IEEE TED 64 (1), pp. 312-318, 2017.

[4] B. Chakrabarti et al., "A multiply-add engine with monolithically integrated 3D memristor crossbar/CMOS hybrid circuit", Nat. Sci. Rep. 7, art. 42429, 2017.

[5] X. Guo et al., "Temperature-insensitive analog vector-by-matrix multiplier based on 55 nm NOR flash memory cells", in: Proc. CICC'17, Austin, TX, Apr.-May 2017, pp. 1-4.

[6] M.R. Mahmoodi and D.B. Strukov, "An ultra low energy internally analog, externally digital vector-matrix multiplier circuit based on NOR flash memory technology", in: Proc. DAC'18, San Francisco, CA, June 2018, art. 22.

[7] M.R. Mahmoodi and D.B. Strukov, "Breaking POps/J barrier with analog multiplier circuits based on nonvolatile memories", in: Proc. ISPLED'18, Bellevue, WA, July 2018, art. 39.

[8] M. Bavandpour, M.R. Mahmoodi, D.B. Strukov, "Energy-efficient time-domain vector-by-matrix multiplier for neurocomputing and beyond", IEEE TCAS-II, 2019 (in print).

[9] M. Bavandpour et al., "Mixed-signal neuromorphic inference accelerators: Recent results and future prospects", in: Proc. IEDM'18, San Francisco, CA, Dec. 2018.

[10] F. Merrikh Bayat et al., "Memristor-based perceptron classifier: Increasing complexity and coping with imperfect hardware", in: Proc. ICCAD'17, Irvine, CA, Nov. 2017, pp. 549-554.

[11] F. Merrikh Bayat et al., "Implementation of multilayer perceptron network with highly uniform passive memristive crossbar circuits", Nat. Comm. 9, art. 2331, 2018.

[12] F. Merrikh Bayat et al., "High-performance mixed-signal neurocomputing with nanoscale floating-gate memory cells", IEEE TNNLS 29, pp. 4782-4790, 2018.

[13] X. Guo et al., "Fast, energy-efficient, robust, and reproducible mixed-signal neuromorphic classifier based on embedded NOR flash memory technology", in: Proc. IEDM'17, San Francisco, CA, Dec. 2017, pp. 6.5.1-6.5.4.

[14] X. Guo et al., "Modeling and experimental demonstration of a Hopfield network analog-to-digital converter with hybrid CMOS/memristor circuits", Frontiers in Neuroscience 9, art. 488, Dec. 2015.

[15] M. Prezioso et al., "Self-adaptive spike-time-dependent plasticity of metal-oxide memristors", Nat. Sci. Rep. 6, art. 21331, 2016.

[16] M. Prezioso et al., "Spike-timing-dependent plasticity learning of coincidence detection with passively integrated memristive circuits", Nat. Comm. 9, art. 5311, 2018.

[17] M.R. Mahmoodi, H. Nili, and D.B. Strukov, "RX-PUF: Low power, dense, reliable, and resilient physically unclonable functions based on analog passive RRAM crossbar arrays", in: Proc. VLSI Symp'18, Honolulu, HI, June 2018, art. 176.

[18] H. Nili et al., "Hardware-intrinsic security primitives enabled by analogue state and nonlinear conductance variations in integrated memristors", Nat. Electron. 1, pp. 197–202, 2018.

[19] S. Sahay, M. Klachko, and D. Strukov, "Hardware security primitive exploiting intrinsic variability in analog behavior of 3D NAND flash memory array", IEEE TED, 2019 (in print).

[20] M.R. Mahmoodi, X. Guo, and D. Strukov, "ChipSecure: A 0.56 pJ/b reconfigurable analog eFlash-based strong PUF with >10210 CRPs and machine learning attack resiliency in 55nm CMOS", in: Proc. DAC'19, Las Vegas, NV, Jun 2019.

Invited Talk

Phase-Change Materials for Storage Class Memory: Novel Bonds at Work

M. Wuttig

RWTH Aachen University of Technology, Germany

Abstract

Phase change media utilize a remarkable property portfolio including the ability to rapidly switch between the amorphous and crystalline state, which differ significantly in their properties. This material combination makes them very attractive as a storage class memory, combing key advantages of Flash and DRAM. This talk will discuss the unique material properties, which characterize phase change materials. In particular, it will be shown that only a rather small group of materials utilizes a unique bonding mechanism ('Bond No. 6'), which can explain many of the characteristic features of crystalline phase change materials. Different pieces of evidence for the existence of this novel bond will be presented. This insight is subsequently employed to develop a novel materials map, which helps to identify systematic property trends and to explore the limits in stoichiometry for such memory applications.

Speaker's Biography

Matthias Wuttig received his Ph.D. in Physics in 1988 from RWTH Aachen/ Forschungszentrum Jülich. From 1995 to 1997 he worked with a Feodor-Lynen stipend at Bell Labs, Murray Hill, New Jersey. He was a visiting professor at several institutions including Lawrence Berkeley Laboratory, Stanford University, Hangzhou University, IBM Almaden, Bell Labs, DSI in Singapore, CiNAM in Marseilles and the Chinese Academy of Sciences in Shanghai. In 1997, he was appointed Full Professor at RWTH Aachen, where his work focusses on the design of novel functional materials. From 2009 to 2018, he was the speaker of the strategy board of RWTH. Since 2011, he heads a collaborative research centre on resistively switching chalcogenides (SFB 917), funded by the German Science Foundation DFG. In 2013, he received an ERC Advanced Grant to realize novel functionalities by disorder control. He is a member of Acatech and the North Rhine-Westphalian Academy of Sciences and has written about 330 publications (~17.000 citations). In 2019 he was selected as an MRS Fellow for path-breaking contributions to the advancement of phase-change materials, including unraveling their unique bonding mechanism, unconventional transport properties and unusual kinetics.

Invited Talk

Bridging Mechanics and Electrochemistry: Experiments and Modeling on Battery Materials

Kejie Zhao

Mechanical Engineering, Purdue University, Bloomington, IN, USA

Abstract

Mechanical failure in batteries is ubiquitous and is well recognized. Less understood is the mechanical behavior of battery materials under chemical load, and the impact of stress on the kinetics of mass transport, charge transfer, interfacial reactions, and hence the potential and capacity of the electrochemical systems. This talk focuses on the interplay of mechanics, such as large deformation, plasticity, and fracture, with chemical reactions in Li-ion batteries. I will discuss the theories of coupled diffusion and stress, stress regulated interfacial reactions, reaction flow, and corrosive fracture in batteries. I will introduce our recently developed operando experiments which probe the continuous evolution of material states during Li reactions and steady degradation of mechanical strength in batteries over cycles. Finite element modeling and atomiostic simulations that emphasize on mechanistic understanding will also be discussed.

Speaker's Biography

Dr. Kejie Zhao joined the faculty of Mechanical Engineering at Purdue University in 2014. He received his PhD degree in Engineering Science in 2012 from Harvard University, and obtained his bachelor's and master's degree from Xi'an Jiaotong University in 2005 and 2008, respectively, He worked as a postdoctoral associate at MIT from 2012 to 2014. The research theme of his group focuses on the chemomechanics of energy materials using experimentations and multi-scale modeling approaches. He is also working on organic semiconductor molecules and polymers and borides chemistry for high-temperature ceramics. He is a recipient of the Extreme Mechanics Letters Yound Investigator Award, 3M Non-tenured Faculty Award, and Haythornthwaite Research Initiation Grant for his research and multiple teaching awards at Purdue University.

Keynote Talk

Nanoelectronics and Heterogeneous Integration with 2D Materials

Eric Pop

Electrical Engineering, Materials Science & Engineering, and SystemX Alliance
Stanford University, Stanford CA 94305, U.S.A. Contact: epop@stanford.edu

Abstract

This talk will present recent highlights from our research on two-dimensional (2D) materials including graphene, boron nitride (h-BN), and transition metal dichalcogenides (TMDs). The results span from material growth and fundamental measurements, to simulations, devices and system-oriented applications that take advantage of unusual 2D material properties. We have grown monolayer 2D semiconductors over large areas, including MoS_2 [1], WSe_2, and $MoSe_2$ [2]. We also uncovered that $ZrSe_2$ and $HfSe_2$ have native high-κ dielectrics ZrO_2 and HfO_2, which are of key technological relevance [3]. Improved electrical contacts [4] led to the realization of 10 nm monolayer MoS_2 transistors with the highest current reported to date, near ballistic limits [5]. These could play a role in 3D heterogeneous integration of nanoelectronics, which presents significant advantages for energy-efficient computation [6]. In less conventional applications, we utilized 2D materials as computing fabrics for analog dot product circuits [7], as thermal insulation for phase-change memory [8], and as the basis of thermal transistors [9]. The latter could enable control of heat in "thermal circuits" analogous with electrical circuits. Combined, these studies reveal fundamental limits and some unusual applications of 2D materials, which take advantage of their unique properties.

Speaker's Biography

Eric Pop is an Associate Professor of Electrical Engineering (EE) and Materials Science & Engineering (by courtesy) at Stanford, where he leads the SystemX Heterogeneous Integration focus area. He was previously on the faculty of UIUC (2007-13) and worked at Intel (2005-07). His research interests are at the intersection of electronics, nanomaterials, and energy. He received his PhD in EE from Stanford (2005) and three degrees from MIT (MEng and BS in EE, BS in Physics). His honors include the Presidential Early Career Award (PECASE), Young Investigator Awards from the Navy, Air Force, NSF and DARPA, and several best paper and best poster awards with his students. He is an Editor of the journal 2D Materials, has served as General Chair of the Device Research Conference, and on program committees of IEDM, VLSI, APS, and MRS. In his spare time, he enjoys snowboarding and electronic music, and he was a DJ at Stanford's KZSU 90.1 FM radio station from 2000-2004. Additional information about the Pop Lab is available online at http://poplab.stanford.edu.

References

[1] K. Smithe et al., ACS Nano 11, 8456 (2017). [2] K. Smithe et al., ACS AMI 1, 572 (2018). [3] M. Mleczko et al., Science Adv. 3, e1700481 (2017). [4] C. English et al., Nano Lett. 16, 3824 (2016). [5] C. English et al., IEDM, Dec 2016. [6] M. Aly et al., Computer 48, 24-33 (2015). [7] N. Wang et al., Symp. VLSI, Jun 2016. [8] C. Neumann et al. in press (2019). [9] A. Sood et al. Nature Comm. 9, 4510 (2018).

Invited Talk

Innovations in Device Packaging: Glass, Lasers, Materials, and Integration

Steve Groothuis

Chief Technology Officer, Samtec Microelectronics in Colorado Springs, CO, USA

Abstract

The use of thin glass wafers has grown exponentially during the past 5 years. Part of that market is the development of glass interposers for high-speed semiconductor package, Lab-on-a-Chip substrate for biomedical applications, electro-optical/photonics packages, and MEMS/Sensors substrate applications. Essential features like Through Glass Via (TGV), laser micromachining, microfluidic channels, and others will be discussed. Samtec Microelectronics has implemented TGVs in glass with integrated Cu interconnects for heterogeneous integration (aka mixed chip integration), biomedical, and sensor applications. Enabling newer device and product technologies (e.g., medical wearables, IoT, and automotive/medical imaging) with glass core technology will be emphasized. Key takeaways will include: understanding the power of implementing glass interposers and substrates, learn about out-of-the-box package engineering, as well as what are the new device packaging opportunities and capabilities.

Speaker's Biography

Mr. Steven Groothuis is the Chief Technology Officer at Samtec Microelectronics in Colorado Springs, CO and maintains the package technology and strategic roadmaps. He is involved in integrated circuit, MEMS, sensor, optical, and photonics package technology pathfinding. He has dealt with products in the autonomous vehicles, biomedical, optical/photonics, MEMS and sensors, and microelectronics arenas. Prior to joining Samtec, he was a DMTS at Micron Technology providing technology advancements on package development, pathfinding, and wafer- and package-related simulations. He also worked at ANSYS as a Multiphysics Industry Specialist on various electronics packaging and MEMS simulation initiatives. He started his career as a package technologist on up to the Advanced Semiconductor Packaging Lab Manager at Texas Instruments. Mr. Groothuis received his BS Physics from Michigan State University and MS Physics from University of Texas at Dallas. He is an IEEE Sr. Member, has coauthored over 50 papers, and holds 14 US patents.

Invited Talk

Emerging Materials for Additive Manufacturing of Flexible Hybrid Sensors

David Estrada

Micron School of Materials Science and Engineering, Center for Advanced Energy Studies, Boise State University, Boise, ID 83725, United States (e-mail: daveestrada@boisestate.edu).

Abstract

Recent advances in the synthesis of 2-dimensional (2D) materials-based inks has increased the design space for additive manufacturing of flexible hybrid electronics (FHE) and sensors. Such systems stand to benefit from high performance flexible silicon IC's for signal processing and amplification, while also leveraging the high surface area and unique physical properties of 2D materials as sensors. For example, previous work has elucidated the fundamental role of crystal defects in the sensing mechanisms of graphene chemiresistors, however, such sensors are yet to find widespread commercial application. Integration of graphene with flexible silicon ICs could help expedite the adoption of such sensors in industry, as sensor response to target analytes could be isolated from other environmental factors through rapid signal processing techniques. Furthermore, advances in additive manufacturing of such systems could further enable widespread adoption of FHE's for applications requiring on-demand manufacturing and repair of sensors and systems. Towards this end, the Advanced Nanomaterials and Manufacturing Laboratory at Boise State University has undertaken several projects to help overcome obstacles facing the integration of 2D materials with FHE systems and sensors. This talk will highlight results of several ongoing studies on the integration of 2D materials with flexible silicon ICs including limiting factors of power dissipation in printed graphene electrodes, the electrochemical response of fully printed graphene electrochemical sensors, and the reliability of flexible silicon die attach strategies for FHE system integration.

Speaker's Biography

Dr. David Estrada is originally from Nampa, Idaho. From 1998 to 2004 he served in the United States Navy as an Electronics Warfare Technician/ Cryptologic Technician – Technical. David achieved the rank of Petty Officer First Class in 2003 before receiving an honorable discharge and returning to Idaho to pursue his undergraduate education at Boise State University (BSU) where he was a Ronald E. McNair scholar. After completing his Bachelor of Science in Electrical Engineering from BSU in May of 2007, he began graduate studies at the University of Illinois at Urbana-Champaign (UIUC) under the direction of Professor Eric Pop; earing his Doctor of Philosophy in Electrical Engineering in 2013. He then returned to Boise State University where he is an Assistant Professor in the Micron School of Materials Science and Engineering. David is the recipient of the NSF, NDSEG, SURGE, and Micron Graduate Fellowships. His work has been recognized with several awards, including the Gregory Stillman, John Bardeen, and Lieutenant General Thomas M. Rienzi graduate research awards, as well at the Society of Hispanic Professional Engineers 2015 Innovator of the Year Award, two AFOSR Summer Faculty Fellowships, and a NSF CAREER Award. His research interests are in the areas of emergent semiconductor nanomaterials and bionanotechnology.

Invited Talk

Phase Transition Materials assisted energy efficient CMOS design

Jaydeep Kulkarni

AMD Chair in Computer Engineering, Assistant Professor of Electrical and Computer Engineering, The University of Texas at Austin, e-mail: jaydeep@austin.utexas.edu

Abstract

Phase Transition Materials (PTM) such as VO2 and NbO2 exhibit abrupt and rapid transition from insulating state to metallic state under the application of electric field. This unique behavior can be leveraged in realizing various energy efficient CMOS circuits. In this seminar, I will present multiple circuit applications of PTM devices. (i) a soft-switching transistor structure (termed as Soft-FET) for voltage droop mitigation in power management applications (ii) PTM devices for in-situ state retention in HD NV-SRAMs (iii) PTM devices for radiation hardened logic and memory applications. I will present preliminary device fabrication and 7nm (predictive) circuit simulation results for the proposed PTM circuit topologies.

Speaker's Biography

Jaydeep Kulkarni is an Assistant Professor in the Department of Electrical and Computer Engineering at the University of Texas at Austin and currently holds AMD endowed chair position in Computer Engineering. He received the Ph.D. degree from Purdue University in 2009. During 2009-2017, he worked as a Senior Staff Research Scientist at Intel's Circuit Research Lab (CRL) in Hillsboro, OR. He has filed 34 patents, published 70 papers in referred journals and conferences. His research is focused on Machine learning hardware accelerators, Memory centric computing, and emerging nano-devices.

Invited Contribution

Invited Contribution

Does CDM ESD Protection Really Work?

Mengfu Di, Han Wang, Feilong Zhang, Cheng Li, Zijin Pan and Albert Wang*

Department of Electrical and Computer Engineering, University of California, Riverside, CA, USA, aw@ece.ucr.edu

Abstract – **On-chip electrostatic discharge (ESD) protection is required to protect ICs against ESD failures of various origins. While ESD protection against human body model (HBM) ESD failures has been very successful, charged device model (CDM) ESD protection design emerges as a new challenge for ICs at advanced technology nodes. This paper reports a comparison study of CDM ESD protection versus HBM ESD protection. It found that the current practices of using conventional pad-based ESD protection networks for CDM ESD protection may not work. This study leads to a game-changing question to the field: Are existing CDM ESD protection approaches fundamentally faulty? It therefore calls for a completely new on-chip CDM ESD protection method to truly resolve the real-world CDM ESD failure problems.**

Index Terms – *ESD protection; CDM; HBM.*

I. INTRODUCTION

There is no question that ESD failure is one of the most devastating IC reliability problems and on-chip ESD protection is required to protect ICs against various ESD damages [1-7]. ESD phenomena have different root causes; accordingly, different ESD protection solutions and ESD protection design characterization methods were developed. Commonly used ESD testing models include HBM model, machine model (MM), CDM model and IEC model. Therefore, various industrial ESD testing standards were developed that have been constantly revised to address the emerging issues of ESD protection for advanced IC technologies [1]. For example, HBM model was initially developed as a military standard [8], which, tough, still serving as the foundation model, has been constantly revised over years, leading to several different industrial HBM standards for varying reasons [9]. On the other hand, as IC technologies continue to advance into sub-32nm nodes and chip complexity continues to increase, CDM ESD model is becoming a major ESD failure challenge [7]. While vast efforts have been devoted to understanding the CDM ESD failure mechanisms and developing CDM ESD protection solutions, it is commonly agreed upon that CDM ESD protection design is still just too challenging today. A few challenging issues with CDM ESD protection, unfortunately, often being mysterious, are: Why are CDM ESD testing results often not repeatable and reproducible? Why does an IC still suffer from CDM ESD failures in field although the chip passed the CDM ESD testing in a factory? Why do CDM ESD failure seem to be so random in field? The fundamental question may be that do the current on-chip CDM ESD protection methods widely used by the industry worldwide really work? This paper reports a study of this fundamental question about CDM ESD protection for ICs.

II. HBM ESD TEST MODEL

It is important to first understand the HBM ESD test model, which has been widely used in the field, largely due to its "reliability" in terms of testing repeatability and reproducibility. Fig. 1a depicts the original HBM ESD waveform defined in the MIL-STD-883E Standard [8], upon which many later-day industrial HBM ESD test models were developed, such as the newest ANSI/ESDA/JEDEC JS-001-2017 [9]. The key parameters for this HBM ESD waveform include the pulse rising time of t_r (<10ns) and decay time of t_d (150ns ± 20ns). Fig. 1b illustrates the equivalent circuit model for the HBM ESD model where a human body accumulates static charges by a charging procedure and ESD discharging occurs when a hand touches an IC part, called a device under test (DUT). Consequently, an HBM ESD transient will zap the IC part, resulting in thermal failure and/or dielectric breakdown to the IC. To address this "external-oriented" ESD discharge risk, each pad on an IC die must be protected by ESD protection structures as illustrated in Fig. 2. In principle, since the HBM ESD surges always come from the "external" world and are

(a)

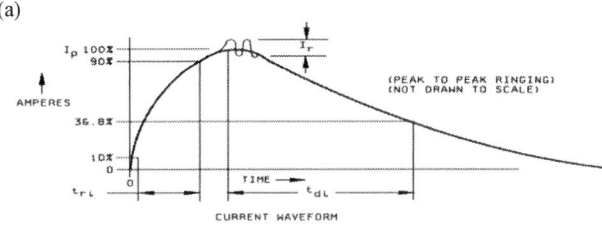

(b)

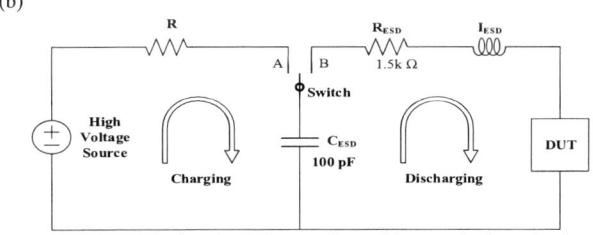

Fig. 1 (a) The original HBM ESD waveform; (b) an equivalent circuit model for HBM ESD charging and discharging procedures [8].

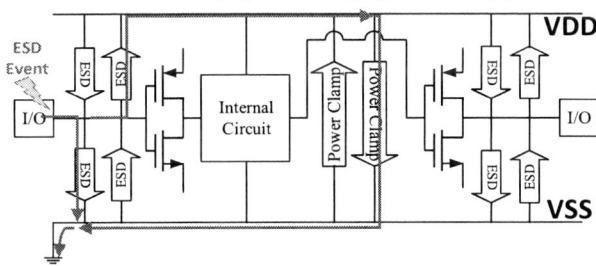

Fig. 2 A classic full-chip ESD protection circuit schematic uses a pad-based ESD protection network on-chip to ensure an ESD discharging path between pads. ESD protection is realized by preventing an ESD pulse from getting inside the IC. This pad-based ESD protection method works for the "from-external-to-internal" ESD discharge models, such as HBM ESD.

injected into an IC at pads, as long as the ESD-critical parameters, including the ESD triggering voltage (V_{t1}), triggering time (t_1) and discharging resistance (R_{ON}), are designed properly for a given IC, when an HBM ESD transient appears at a given pad, the ESD protection structure (s) connected to this pad will be turned on promptly to discharge the ESD pulse, i.e., an ESD transient will be bypassed externally and will not flow into the IC, hence protecting the IC. Due to the fact that the HBM ESD phenomenon is well understood and the HBM ESD waveforms are relatively slow, HBM ESD protection designs have been fairly successful today in the industry. To be clear, the distinct feature for HBM ESD model, as well as MM and IEC models, is that the ESD pulse comes from the "external" world, which can be blocked from getting into an IC chip using the ESD protection structures connected at the pads. This is, however, not the case for CDM ESD phenomena.

III. CDM ESD TEST MODEL

CDM ESD phenomenon is entirely different from HBM, MM and IEC ESD models. As illustrated in Fig. 3, CDM ESD model features a self-charging/discharging procedure that models a real-world ESD phenomenon where an IC (or broadly, an electronic component) is charged by various possible procedures, triboelectrically or by electrostatic field induction, during its life time and accumulates substantial electrostatic charges inside. If the IC is then grounded, the charges stored inside will discharge into the ground. The resulting CDM ESD transient flowing from inside of the IC to the ground may cause thermal or dielectric failures to the IC [1]. Compared to HBM ESD model, the CDM ESD pulse is very short (full width at half maximum, or, FWHM = 250~600ps) and rises extremely fast (t_r <250ps), as depicted in Fig. 4 [10]. For example, the newest CDM ESD standard (ANSI/ESDA/JEDEC JS-002-2014) describes the device level CDM ESD testing procedures as shown in Fig. 4, where the DUT is charged through electrostatic field induction, and the static charges generated and stored inside the DUT will be discharged into the ground when the pogo pin moves close to and/or touches the DUT later. The CDM ESD tester models the real-world CDM ESD charging and discharging procedures. As the DUT, an IC can be a bare die or a packaged device. The Ground may be a printed circuit board (PCB) on which the IC is mounted. CDM ESD model has several unique features: First, the waveform is oscillatory with very short rise time and duration. Second, the waveform is very sensitive to the tester set-up and, in case of testing a packaged IC, to the packaging parasitic parameters.

(a)

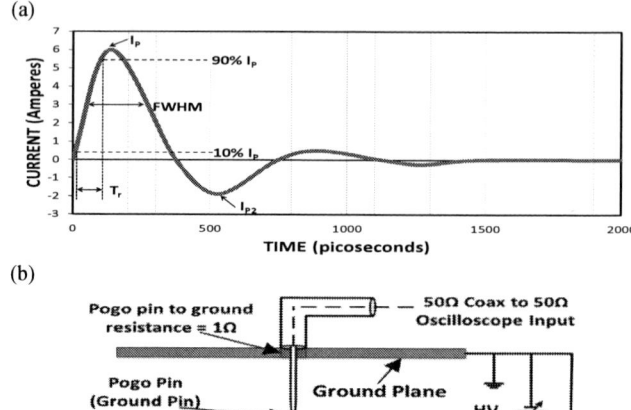

(b)

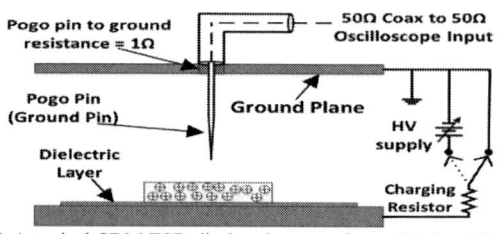

Fig. 4 (a) A typical CDM ESD discharging waveform; (b) simplified CDM ESD test model schematic [10].

Third and very critically, the CDM ESD phenomenon involves an ESD transient flowing from the "inside" of an IC to "external" ground. Hence, the "from-internal-to-external" CDM ESD model is entirely differently from other "from-external-to-internal" ESD discharging procedures for HBM, MM and IEC ESD models. Notoriously, CDM ESD testing results are poor in repeatability and reproducibility, and CDM ESD protection often fails randomly in field regardless of the in-lab testing results. Consequently, CDM ESD protection is becoming a real pain to IC designers today. We believe that the CDM ESD design challenge is not entirely due to the fast and sensitive natures of the CDM ESD model. We found that the root cause of today's CDM ESD design problem may be associated with the conventional pad-based CDM ESD protection method, commonly used in the industry today, which unfortunately may be fundamentally faulty.

IV. CDM ESD PROTECTION CASE STUDY

An IC as a DUT can be a bare Silicon die or a packaged IC as illustrated in Fig. 5. Both a bare Si die (wafer level) and a packaged IC part will suffer from CDM ESD damages in real world [2, 11]. While the focus has been with the packaged ICs in early years, the industry is getting more and more concerned with the bare die CDM ESD damages, i.e., wafer level CDM ESD failures, recently. Theoretically, CDM ESD failures can occur at any level, from bare dies to package to modules. Hence, adequate CDM ESD protection must be designed to protect ICs at any level. Ironically, while it is the common understanding that CDM and HBM ESD models are sharply different, it is a

(a)   (b)

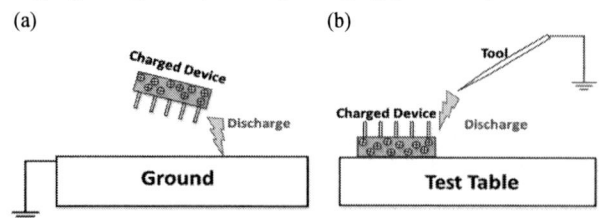

Fig. 3 Left: In a real world, an IC part may be charged by various procedures. The static charged stored inside will be discharged when the IC is grounded, resulting in CDM ESD failures. Right: A CDM ESD tester schematic for CDM discharging.

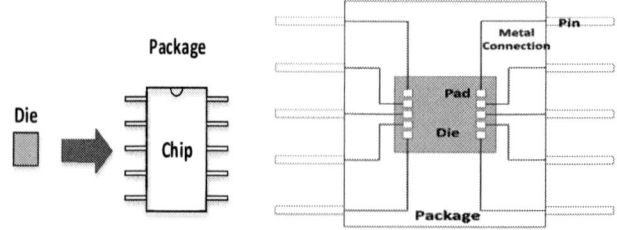

Fig. 5 IC as DUT: From Left to Right, a Si die is packaged with the package internal schametic depicted.

common practice today in the industry that on-chip CDM ESD protection still follows the same approach as used for HBM ESD protection, i.e., ESD protection structures are connected to the IC pads (Fig. 2). Typically, engineers believe that the focus on CDM ESD protection designs should be to address the ultrafast and short natures of CDM ESD pulses as compared to slower HBM ESD surges. The response time of a CDM ESD protection structure to an ultrafast CDM surge is certainly a critical design consideration. However, we believe that it is the existing pad-based CDM ESD protection network method that may be the real devil making CDM ESD design challenging today, as discussed below.

A. CDM ESD protection for packaged ICs

The common pad-based ESD protection scheme as depicted in Fig. 2 ensures that ESD protection structures are connected to every pad and there must be an ESD discharging path between every two pads on an IC in order to guarantee a low-R ESD discharging path for any possible ESD stressing scenarios [1]. This is proven to be successful for HBM ESD protection due to its "from-external-to0internal" nature. For a packaged IC device, this full-chip ESD protection scheme is illustrated in Fig. 6 where the central ESD block represents a pad-based ESD protection network connected to the pads to ensure an ESD discharge path between pads. For a charged packaged IC part, it is generally believed that the static charges shall be stored on the package frame, and the concerned electrostatic charges shall stay on the metal interconnects (e.g., supply buses) [7]. When a CDM ESD event occurs, i.e., one package pin is suddenly grounded, as depicted in Fig. 6, the accumulated static charges on the package will, theoretically, always be able find a way to flow to the grounded pin to discharge into the ground, either directly through the package metal interconnects if located nearby, or, through the ESD protection structures connected to the pads on the die, then to the grounded pin. This means that the accumulated charges will not run through the internal die, hence, in principle, CDM ESD protection can be realized by the ESD protection structures connected at the pads. In the micro scope, i.e., within the package, the CDM ESD discharge can still be considered being "from external to internal" and these charges will be kept away from the internal die by the ESD protection structures at the pads. However, this ideal assumption will fail in many cases. In

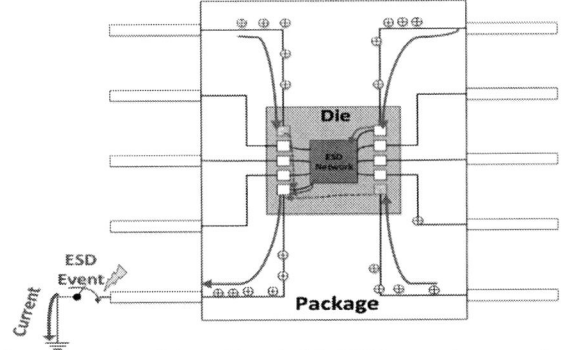

Fig. 6 Illustration of the pad-based global ESD protectio nnetwork on an IC chip where ESD protection devices are connected to each pad to ensure an ESD discharging path between pads on a chip.

a real world, this pad-based CDM ESD protection mechanism may not work. For example, while an ESD device may be designed working in a given discharging mode, the ESD-critical parameters (t_1, V_{t1} & R_{ON}, etc.) may not work in both directions, and/or as in combination with other devices on a die, when dealing with the charges may stay everywhere on a package frame. This means that, in certain cases and likely randomly, one or more ESD protection devices in a discharge path may not be triggered properly, and consequently, the charges that are supposedly discharged through this path may be deviated into the inside of the IC die, which may result in ESD damages to internal transistors, hence leading to CDM ESD failures. In a second example, for advanced ICs featuring very high frequency, ultra-wide bandwidth and extremely high data rates, to avoid the unbearable negative impacts of the ESD-induced parasitic capacitance on the IC performance, some critical signal pads are often not protected by ESD devices, which apparently breaks the global ESD protection network of the packaged IC. Consequently, some stored charges may have to flow through the inside of the IC dies in order to discharge through the ground pin on a package, hence results in CDM ESD failure internally, as illustrated in Fig. 7. In summary, the conventional pad-based ESD protection scheme works for the "from-external-to-internal" ESD discharge models, such as, HBM, MM and IEC models; however, in principle, it will not work for the "from-internal-to-external" CDM ESD model.

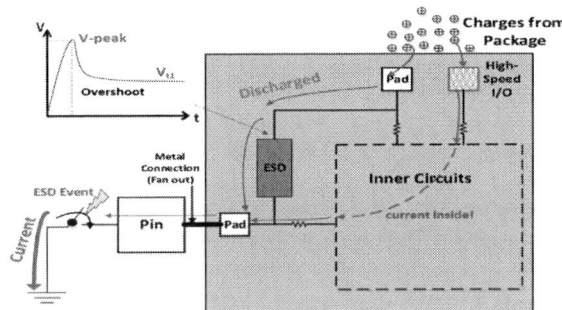

Fig. 7 Illustration of ESD discharge of charges stored on a package for an IC with certain data-critical pads not protected by ESD devices.

B. CDM ESD protection for bare IC dies

In case of wafer level CDM ESD phenomena occurring to bare Si dies without packaging, the CDM ESD discharging procedures will be very different from the packaged IC cases. As depicted in Fig. 8, for a bare Si die, the static charges generated by whatever procedures will be stored inside the IC die throughout, possibly in the Si substrate, along the metal interconnects or anywhere inside the die [2, 11]. Assume that the pad-based ESD protection scheme is used as shown in Fig. 8, when one pad is grounded and CDM ESD discharge then occurs. For the charges that stored nearby the ground pad (e.g., V_{SS}), it is possible that those charges can be discharged into the ground through the ESD protection device connected to the nearby pad, hence, CDM ESD protection can be realized. However, for a large and complex chip, the large probability may be that substantial charges stored everywhere must find their way to flow to the grounded pad to discharge into the ground. Therefore, these charges will have to flow through the

(a)

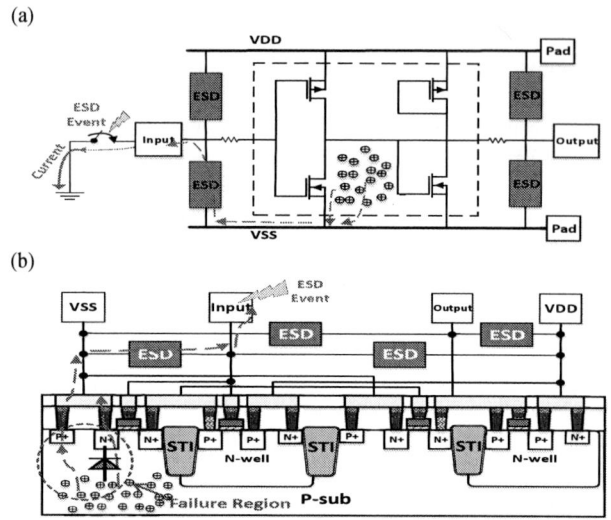

(b)

Fig. 8 Schematic and cross-sectional illustration of the problem with the traditional pad-based global CDM ESD protection scheme.

internal IC die, randomly and unpredictably, even if they would find their ways to the ground. This means that substantial CDM ESD discharge currents will flow through some internal transistors, which will cause thermal or dielectric failures inside an IC; hence, CDM ESD damages will occur even though there ESD protection devices connected to each pads. Further, even if these ESD protection were designed wonderfully and verified individually in CDM ESD testing, it will possibly never be able to protect the IC against the CDM ESD surges as expected. Fig. 9 shows a case where, during an CDM ESD event, the locally stored charges cannot flow through the reverse PN junctions, hence quickly raises the gate voltage locally and causes gate breakdown internally. The root cause of the above discussed CDM ESD failure cases is associated with the pad-based global ESD protection principle that has been commonly used today. Though, this pad-based global ESD protection network method will work for the "from-external-to-internal" ESD discharging events, including HBM, MM and IEC models, by stopping the external ESD pulses from getting inside of an ICs; it in principle will fail the "from-internal-to-external" CDM ESD model where, in most cases, the ESD discharge current will have to flow through the internal IC die, and hence, CDM ESD failures will likely to happen even with ESD protection devices connected to all pads. Therefore, we believe that the existing

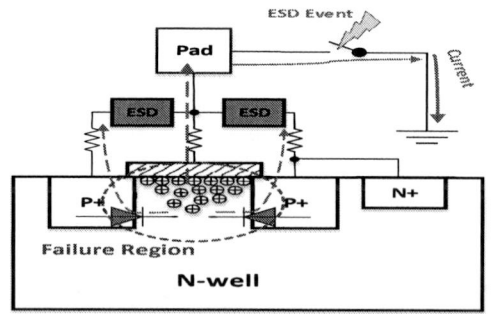

Fig. 9 Illustration for possible CDM ESD failure inside a PMOS FET where the locally stored charges cause high potential across the gate during CDM ESD discharging, resulting in gate breakdown.

pad-based CDM ESD protection method commonly used in the industry today is fundamentally faulty. It is fair to say that if the CDM ESD protection works for an IC, it is purely lucky. This unavoidable and often random "from-internal-to-external" CDM ESD failure model may be the root cause to the CDM ESD design challenges faced by the industry today, including lack of repeatability and reproducibility, as well as random CDM ESD failures in field. We propose a novel paradigm-changing full-chip CDM ESD protection method to overcome this fundamental CDM ESD design challenge, which will be reported elsewhere.

V. CONCLUSION

In summary, this paper discusses the mechanism of CDM ESD phenomena on IC chips. It is found that the traditional pad-based global CDM ESD protection network method, commonly used in the industry today, may be fundamentally faulty due to the "from-internal-to-external" nature of CDM ESD discharge procedures. It is believed that the erroneous pad-based CDM ESD protection approach causes the CDM ESD failure uncertainties frequently observed today. Therefore, an entirely new full-chip CDM ESD protection method must be developed for real-world CDM ESD protection. In short, our words to IC designers are clear: your beloved pad-based CDM ESD protection methods may be fundamentally faulty!

REFERENCES

[1] A. Wang, *On-Chip ESD Protection for Integrated Circuits: an IC Design Perspective*, Kluwer, Boston, 2002, ISBN: 0-7923-7647-1.

[2] S. Voldman, *ESD Testing: From Components to Systems,* Wiley & Sons, 2017.

[3] A. Wang, et al, "A Review on RF ESD Protection Design", *IEEE Trans. Electron Devices, Vol. 52, No. 7*, pp. 1304-1311, July 2005.

[4] H. Feng, et al, "A Mixed-Mode ESD Protection Circuit Simulation-Design Methodology", *IEEE J. Solid-State Circuits, V38, No. 6*, pp. 995-1006, June 2003.

[5] F. Lu, et al, "A Systematic Study of ESD Protection Co-Design with High-Speed and High-Frequency ICs in 28nm CMOS", *IEEE Trans. Circuits and Systems I: Regular Papers, Vol. 63, No. 10*, pp.1746-1757, October, 2016.

[6] F. Zhang, et al, "A Full-Chip ESD Protection Circuit Simulation and Fast Dynamic Checking Method Using SPICE and ESD Behavior Models", *IEEE Trans. on Computer-Aided Design of ICs and Systems, 2018.* DOI: 10.1109/TCAD.2018.2818707.

[7] H. Wang, et al, "Chip-Level CDM Circuit Modeling and Simulation for ESD Protection Design in 28nm CMOS", *Proc. IEEE ICSICT, 2018*, DOI: 10.1109/ICSICT.2018.8564936.

[8] "Electrostatic Discharge Sensitivity Classification", Method 3015.7, MIL-STD-883E, Dept. of Defence, March 1989.

[9] "For Electrostatic Discharge Sensitivity Testing, Human Body Model (HBM) - Component Level", *ESDA and JEDEC*, ANSI/ESDA/JEDEC JS-001-2017, ESD Association and JEDEC Solid State Technology Association.

[10] "For Electrostatic Discharge Sensitivity Testing, Charged Device Model Model (CDM) - Device Level", *ESDA and JEDEC*, ANSI/ESDA/JEDEC JS-002-2014, ESD Association and JEDEC Solid State Technology Association.

[11] N. Jack, "Charged Device Model ESD Protection and Test Methods for ICs", a PhD Dissertation, University of Illinois at Urbana-Champaign, 2012.

Contributed Paper Sessions

Contributed Paper

Numerical Study of Non-Circular Pillar Effect in 3D-NAND Flash Memory Cells

Albert Fayrushin, Haitao Liu, Aurelio Mauri, Gianpietro Carnevale, Hyejin Cho, Duo Mao

Abstract—Two-dimensional (2D) Technology-Aided Computer Design (TCAD) simulation method is proposed for evaluation of program/erase/retention degradation of Three-dimensional (3D) NAND Flash memory induced by structure deformation. Incremental step-pulse program (PGM) and erase (ERS) electrical characteristics are calculated for various spike-like deformations of vertical channel etch hole, including magnitude, width and amount. Strong dependence of erase speed and negative charge retention on spike deformation is found.

Index Terms—Flash memories, Semiconductor device modeling.

I. INTRODUCTION

SINCE the first commercial introduction of 3D NAND Flash [1-2] the production volume of vertical NAND Flash memories has been continuously expanded, resulting in recent revenue parity with planar 2D NAND Flash [3]. 3D NAND Flash advantages in cost-per-bit reduction are accompanied with increased structural complexity and new process challenges [4]. One of such challenges is to maintain equal size and shape of high-aspect ratio etch holes made before deposition of the memory cell stack [5]. A typical issue is the difficulty to obtain a perfect circular cross section during the etch hole process: elliptical and spike-wise distortions cross-sections [6]. To enable 4-bit per cell operation, it is important to predict and mitigate the effects induced by etch hole deformations on program-erase performance and charge retention. The purpose of this work is the analysis and correlation of the above mentioned geometrical deformations with the programming, erase, retention electrical signatures by means of TCAD modeling.

II. MODEL

Transmission electron microscopy (TEM) image of NAND Flash memory at Fig.1 [6] is an example of possible etch hole deformations. Elliptical and spike-like types can be defined (Fig.2). The structures of Fig.2 are realized by implementing into TCAD simulator [7] spike shape statistically controlled with the help of a Normal distribution (1).

A. Fayrushin, H. Liu, D. Mao are with the Micron Technology, Boise, ID 83716 USA (e-mail: afayrushin@ micron.com). H. Cho is with the Micron

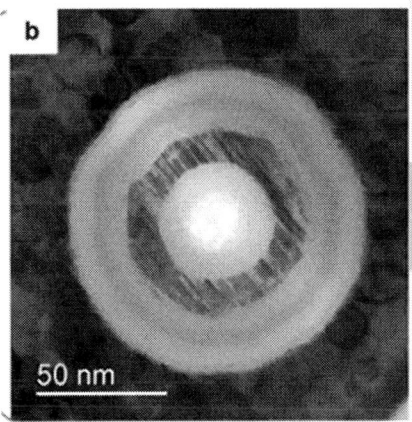

Fig.1. Example of etch hole deformations (printed with permission of JFE Techno-Research Corporation, www.jfe-tec.co.jp [6]).

$$f = A \cdot e^{-\frac{(x-\mu)^2}{2\sigma^2}}, \qquad (1)$$

where μ is mean of expectation (here 0), σ is the standard deviation, A – magnitude.

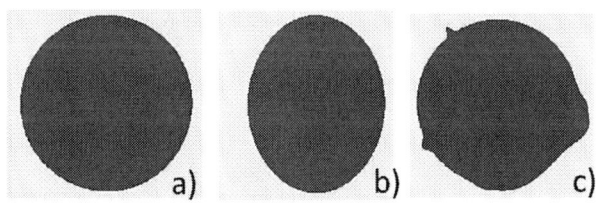

Fig.2 Etch hole shapes for 3D NAND Flash: a) non-deformed; b) ellipsoidal; c) with spikes of different width.

Pillar deformation is a spatial defect, whose careful implementation in 3D framework requires a very large number of mesh points. To avoid potential numerical issues and to deal with an acceptable simulation time a planar 2D modeling approach is employed. To verify the correctness of this approximation a comparison is made with full 3D approach in the case of non-deformed cells. A 3D model (Fig.3) is realized starting from Fig.1 adjusting the stack layers to obtain reasonable PGM/ERS performance (Fig.4).

Technology, Singapore. A. Mauri, G. Carnevale are with the Micron Technology, Agrate, Italy.

978-1-5386-1202-6/19 $31.00 © 2019 IEEE

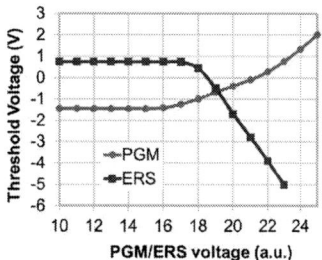

Fig.3. Full 3D model of memory cell [6].

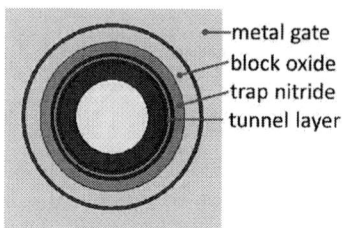

Fig.6. Non-deformed cross-section model of 3D NAND cell.

Fig.4. ISPP/ISPE curves of full 3D cell model.

In the 2D-planar framework the main issue is related to the threshold voltage simulation since channel is in the normal direction. To solve this issue a correlation between channel current and channel conductivity is obtained using the 3D model. Fig.5 shows IDVG (drain current - gate voltage) curves obtained with different amount of electrical charge in trapping layer and compared with the average channel conductivity extracted at the middle of the cell (half of gate length position). Conductivity is obtained for the condition of grounded drain/source. A good correlation, as expected, is obtained between conductivity and drain current: instead of IDVG simulation we can estimate the cell threshold variation just by conductivity calculation.

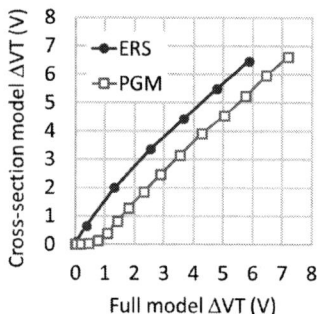

Fig.7 Absolute change of threshold voltage during PGM/ERS for full 3D model versus cross-section model for the same PGM/ERS bias.

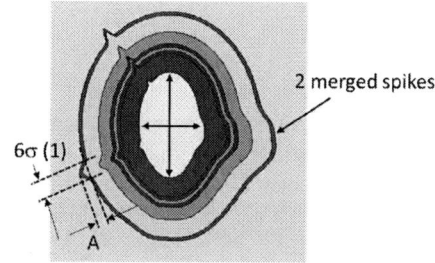

Fig.8 Example of cross-sectional model with applied spiked and ellipsoidal deformations

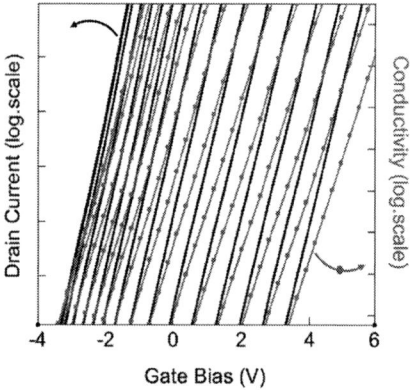

Fig.5 Current-Gate Voltage characteristics and Conductivity-Gate Voltage during PGM operation.

The latter calculation can be easily realized also in the 2D framework. Fig.6 shows the 2D cross-section of the non-deformed case obtained with same stack composition and dimensions of the model in 3D (Fig.3). Another challenge in the 2D-planar framework is the modeling of programming/erase operations that determines the distribution of Nitride trapped charge due to the possible lateral charge spreading and gate corner effects not taken into account in 2D.

This is realized by adjusting the tunneling parameters: doing that similar ISPP/ISPE curves between cross-section model and fully 3D model are obtained as shown on Fig.7 (threshold voltage variation is calculated with the conductivity model). Curves are not linear due to the absence of lateral spread in 2D model.

Fig 8. shows how by means of Equation (1) circular shaped cross-section can be distorted, adding desired number of spikes with arbitrary width and height

III. RESULTS

A. Single spike deformation

Initial analysis is performed considering a single deformation spike with variation on spike extension and width. Fig.9 shows generated structures having fixed spike extension of 9 nm and variable spike width. As spike gets wider the shape transforms from circular to ellipsoidal. Simulated PGM and ERS characteristics are presented on Fig.10. To estimate how PGM/ERS speed is affected by etch hole deformation we calculate the bias required to achieve a threshold variation of 4V PGM and 2V for ERS.

978-1-5386-1202-6/19 $31.00 © 2019 IEEE

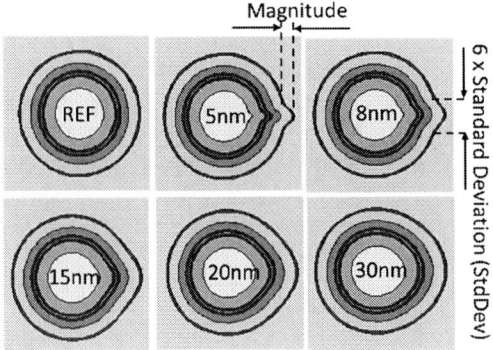

Fig.9 Generated pillar cross-sections having fixed deformation spike magnitude and variable spike width (standard deviation is denoted by numbers).

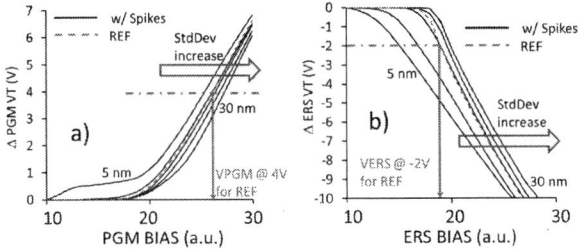

Fig.10 Dependence of threshold voltage shift on Program (a) and Erase(b) gate bias for spike standard deviation (StdDev): 5/8/15/20/30 nm and spike magnitude of 9 nm.

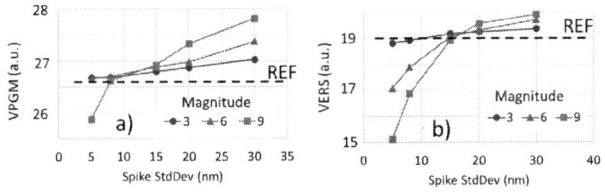

Fig.11 Program speed (a) and erase speed (b) dependence on spike magnitude and width.

Fig.11 reports this estimation as a function of VPGM and VERS, which are monotonically rising with spike magnitude increase and standard deviation decrease, crossing values for non-deformed pillar (REF comes from Fig.10). Presence of narrow spike leads to acceleration of local program speed at the center of the spike and to the retarded program speed between spike and non-deformed region as shown in Fig.12. As a result, trapped charge distribution around the spike becomes non-uniform, enhanced at spike center and depleted outside the spike. At high spike amplitude (9 nm) and minimum sigma (5 nm) increased local trapped charge leads to overall program speed enhancement (VPGM drop). Spike width increase results in reduction of local electron concentration and reduction of program speed (increase of VPGM). At some standard deviation value trapped charge depletion compensates local charge enhancement and VPGM becomes equal to VPGM of non-distorted cell. Further increase of spike width leads to the effect of local etch hole radius increase (StdDev 20-30 nm) and program speed becomes less than for the reference case.

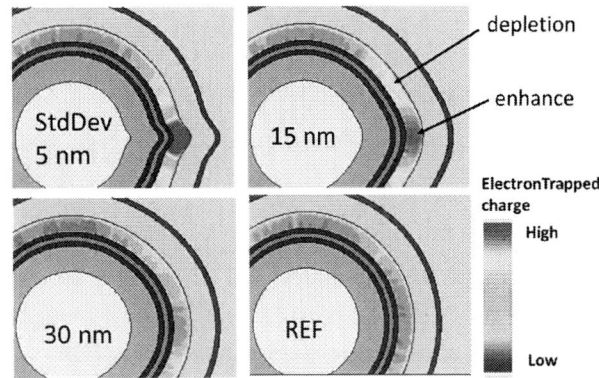

Fig.12 Electron trapped charge distribution, extracted at gate bias equal to VPGM, in storage Nitride for the reference and distorted (spike magnitude 6 nm) cases.

Erase charge distribution in storage nitride depending on spike width is similar to program operation but has higher sensitivity to deformation spikes as VERS is mainly determined by the charge under deformation spike (Fig.13) via conductivity concentration. Magnitude of the spike amplifies spike influence on VPGM/VERS (Fig.11). Interesting property is existence of the spike standard deviation when VPGM and VERS do not depend on spike magnitude.

We also simulated the impact of the etch hole deformation on the number of lost electrons from a program state: fig.14 reports this number (at fixed retention time and temperature) calculated among the different structures. Initial number of electrons corresponds to VPGM. Charge loss is accelerated at narrow spike width and higher spike magnitude, exceeding reference (non-deformed cross-section) values. As in the case of program and erase, retention is aggravated by the enhancement of the electric field in the deformation spike (Fig.15). For the very wide spikes retention is improved as local etch hole radius increases.

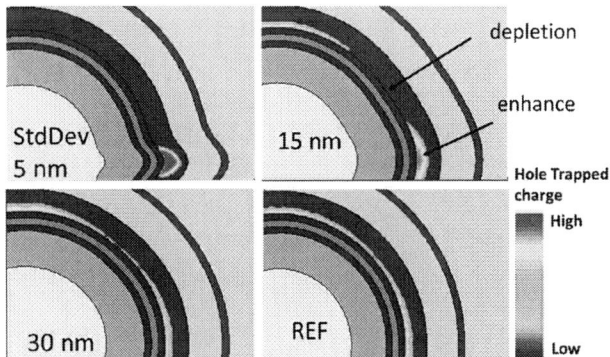

Fig.13 Hole trapped charge distribution, extracted at gate bias equal to VERS, in storage Nitride for the reference and distorted (spike magnitude 6 nm) cases.

B. Multiple spike deformation

In this section we consider the effect of multiple spikes. To do that, multiple spikes are added to each of the StdDev/Magnitude split condition (Fig.16) obtaining a total of 45 samples (3 magnitudes x 5 StdDev's x 3 spike numbers).

978-1-5386-1202-6/19 $31.00 © 2019 IEEE

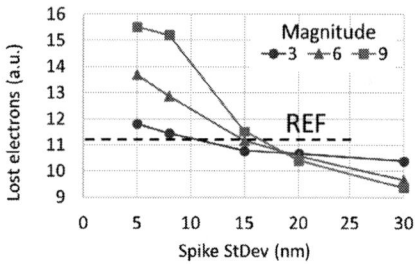

Fig.14 Number of electrons left storage nitride at room temperature and fixed period of time (~ days).

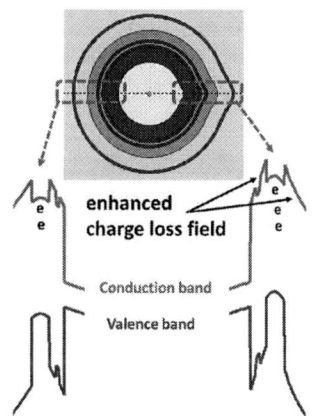

Fig.15 Band diagram showing reduced charge loss barrier for spike region (StDev 8 nm, Magnitude 6 nm).

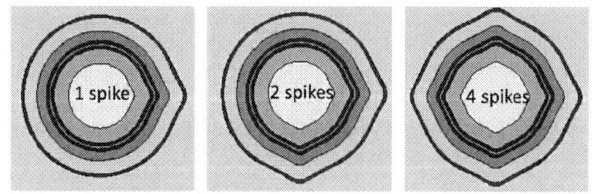

Fig.16 Extension of simulation samples adding structure with 2 and 4 spikes

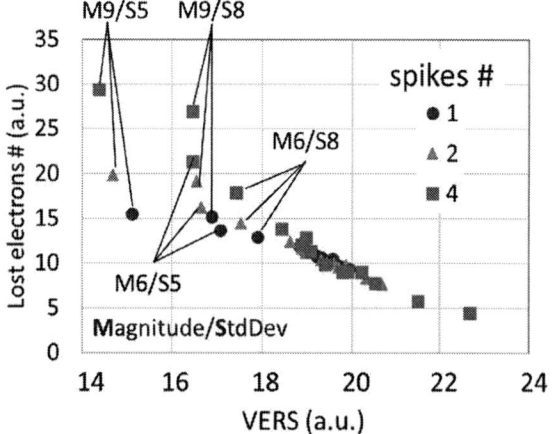

Fig.17 Reverse linear correlation between lost electrons at retention and erase speed indicator (VERS).

Reverse linear correlation is found between number of lost

electrons and VERS in the range of standard deviation 15-30 nm (Fig.17). At lower spike width values correlation line gets wider for retention variable driven by number of spikes in the cell cross-section. For narrow spikes VERS is less dependent on spikes number than retention. Erase speed and electronic charge loss are correlated through the process of removing electrons from storage nitride layer. No straight correlation is found between lost electrons and program speed as shown on Fig.18.

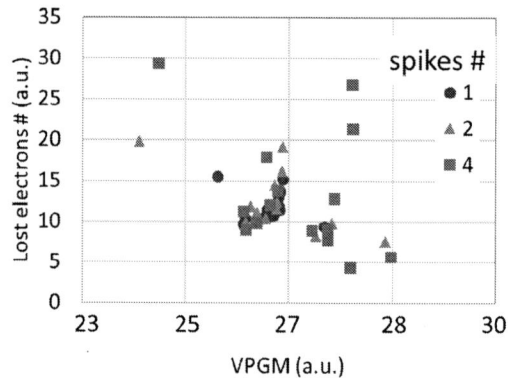

Fig.18 Correlation plot between lost electrons at retention and program speed indicator (VPGM).

IV. CONCLUSION

TCAD simulations are used to correlate the effects of non-circular channel spike deformations with electrical characteristics of the 3D Nitride NAND Flash memories. Such deformations result in higher impact on erase speed rather than on program speed and to accelerated program charge loss. This correlation can be employed to detect the non-circular and distorted shapes without any physical analysis. As consequence process control of etch hole geometry should be of importance to minimize pillar to pillar variations and enable transition to 4-bit per cell operation.

REFERENCES

[1] K.-T. Park et al., "Three-dimensional 128 Gb MLC vertical NAND flash-memory with 24-WL stacked layers and 50 MB/s high-speed programming", *IEEE ISSCC Dig. Tech. Papers*, pp. 334-335, Feb. 2014.

[2] K. Parat, C. Dennison, "A floating gate based 3D NAND technology with CMOS under array", *IEDM Tech. Dig.*, pp. 48-51, Dec. 2015.

[3] H.H. Jones, "Memory Technology and Overall Trends in the Semiconductor Industry," presented at *Semicon Taiwan*, Zhubei City, Hsinchu County, Taiwan, Sept. 7-9, 2016.

[4] V. Vahedi, "Etch Challenges and Solutions," presented at *Semicon China*, Shanghai, China, March 15-17, 2016.

[5] J. Jang et al., "Vertical Cell Array using TCAT (Terabit Cell Array Transistor) Technology for Ultra High Density NAND Flash Memory", Dig. Symp. VLSI Tech., pp. 192-193, June 2009

[6] Cross-sectional TEM analysis of micro devices. [Online]. https://www.jfe-tec.co.jp/analysis/device01.html

[7] "Sentaurus Device Simulator", Synopsys Inc, 2018.

978-1-5386-1202-6/19 $31.00 © 2019 IEEE

Contributed Paper

Metal-Site Dopants in Two-Dimensional Transition Metal Dichalcogenides

I. Williamson, M. Lawson, S. Li, Y. Chen, and L. Li

Abstract— Two-dimensional transition metal dichalcogenides (2D TMDs), have received a lot of attention for having notable structural, electrical and optical properties. 2D TMDs can be further tuned through the implementation of substitutional dopants. This work utilizes density functional theory in order to screen the effects of six different transition metal-site dopants (Mo, Ni, Sc, Ti, V, and W) on the structural and electrical properties of 2D MX_2 (M = Mo and W; X = S, Se, and Te). Dopant stability was found to be largely dependent on the atomic radii of the dopant and host metal atoms. Electronic band gap calculations reveal W-doped MoX_2 and Mo-doped WX_2 to be among the only observed semiconductors. The photosensitivity and photoresponsivity are significantly enhanced by doping with Sc or Ni. This work offers an extensive investigation of metal-site dopants and their impact on 2D MX_2 materials.

Index Terms— 2D semiconducting materials, dopants, computational screening, first-principles, density functional theory

I. INTRODUCTION

Over the past decade, there has been growing interest into the field of two-dimensional nanostructures due to their novel properties. Transition metal dichalcogenide (TMD) monolayers have potential to replace graphene as the most promising candidate for next-generation device applications because they offer a variety of compositions, many of which have a band gap. These binary compounds have the chemical formula MX_2 (M = transition metal, X = chalcogen) where M atoms reside between two X layers forming a monolayer sheet (see Fig. 1). Each 2H-type MX_2 sheet has D6h point group symmetry. Similar to graphene, strong covalent bonds hold the sheets together while primarily weak van der Waals forces exist between sheets. This allows for mechanical exfoliation of the sheets.[1] 2D TMDs have interesting electric,[2-4] thermoelectric,[5,6] and photoelectric properties[7] which remain somewhat untapped

I. Williamson was with the Micron School of Materials Science and Engineering, Boise State University, Boise, ID 83725, USA. He is now with Micron Technology, Inc. (email: iwilliamson@micron.com).

M. Lawson is now with the Micron School of Materials Science and Engineering, Boise State University, Boise, ID 83725, USA (e-mail: matthewlawson@u.boisestate.edu).

S. Li and Y. Chen are both now with Department of Mechanical Engineering, The University of Hong Kong, Hong Kong SAR, China (emails: shashali@connect.hku.hk and yuechen@hku.hk).

L. Li is now with the Micron School of Materials Science and Engineering, Boise State University, Boise, ID 83725, USA (e-mail: lanli@boisestate.edu).

due to the large range of composition and structure possibilities. Recent studies have been conducted to narrow the scope and potentially lead to novel 2D TMD materials.[8-11]

Transitioning from bulk to monolayer structures modifies the properties of the TMDs. For example, 2D MoS_2 has shown promising benefits as a channel material for field effect transistor (FET) applications due to its reduced dimensionality.[3, 12] 2D MoS_2 also experiences a shift from indirect (in bulk) to direct band gap semiconducting behavior which is useful for potential electronic and optoelectronic applications.[13] It was found that the photoluminescence of MoS_2 increases as its thickness decreases,[14] as a result, the luminescence quantum efficiency of monolayer MoS_2 is four orders of magnitude stronger than that of bulk MoS_2.[15]

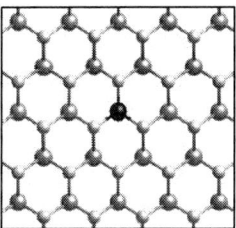

Fig. 1. Top-down view of M'-doped MX_2 where the M' dopant atom is depicted with a black sphere, the M atoms are gray spheres, and the X chalcogens are yellow spheres.

Further property improvements and tuning can be achieved through the use of substitutional dopants on metal or chalcogenide sites, yielding interesting changes in the intrinsic properties of 2D TMDs[11]. This paper focuses on metal-site dopants, because they cause enhancements such as inducing and tuning the magnetic properties.[16,17] They also offer the ability to modify the band gap and photoluminescence[18] and control whether it is a p- or n-type semiconductor.[19] Previous work has shown that these dopants are stable in 2D TMDs and that they are a substitutional dopant rather than an interstitial impurity.[20,21] This type of defect tends to be more stable because the dopant inherits the strong covalent bonds within the TMD sheet.

In this work, we investigate the effect of six different transition metals (Mo, Ni, Sc, Ti, V, and W) as metal-site dopants in 2D-type MoX_2 and WX2 (X = S, Se, or Te). Both MoX_2 and WX_2 are most stable in the 2D-type structure. The six transition metals were chosen based on the screening criteria of previous work[9] which incorporated the transition metal criticality scores of Graedel *et al.*[22] The "criticality" of transition metals in this context refers to their estimated supply

978-1-5386-1202-6/19 $31.00 © 2019 IEEE

risk, environmental impact, and vulnerability to supply restriction. This amounts to a total of 30 doped structures investigated. The structural and electrical properties were analyzed with respect to the physical characteristics of the dopant atoms (e.g., atomic radius, preferred oxidation state) to better understand the doping effects of these 2D MX_2 structures. This approach offers a means of extensively analyzing the effects of dopants in MoX_2 and WX_2 through a broad range of transition metal dopants to identify key factors in improving novel 2D TMD materials. Onofrio *et al.* performed first-principles calculations to study substitutional and interstitial doping of Mo and W dichalcogonides to determine the trends in energetics and electronic properties[11]. Our paper focuses on Mo, Ni, Sc, Ti, V, and W substitutional dopants to explore their potential electronic and optoelectronic applications.

II. COMPUTATIONAL METHODS

Ground state calculations were conducted using the Vienna *ab-initio* simulation package (VASP) code within the framework of density functional theory (DFT).[23] The spin-polarized generalized gradient approximation (GGA) was employed using the Perdew Burke Ernzerhof (PBE) exchange correlation functional with a 500 eV plane-wave cutoff energy.[24] Projector-augmented wave (PAW)[21] pseudopotentials were used along with a Gaussian smearing of 0.05 eV. Brillouin zone integration was performed over a Γ-centered 6x6x1 k-point mesh. To account for the metal-site dopant, each structure was generated as a 4x4 supercell of the corresponding MX_2 where single-layers were separated by a vacuum of at least 20 Å. By replacing a single metal in each cell with a dopant atom, the resulting chemical formulas were of the form $M'M_{15}X_{32}$ (where M' = dopant atom Mo, Ni, Sc, Ti, V, or W; M = Mo or W; and X = S, Se, or Te). Each structure was relaxed until residual forces were within 0.01 eV/Å and stress in the periodic direction was less than 0.03 GPa. We only considered MX_2 monolayer without interlayer interaction, so we did not include van der Waals effect in the calculations for computational efficiency.

For electronic structure calculations, a denser 24x24x1 k-point mesh was used. Strong correlation effects from the transition metals were accounted for by the implementation of DFT+U.[23] In this approach, convergence-tested on-site Coulomb potentials of U = 4.38, 8.0, 3.0, 8.0, 6.0, and 8.0 eV were used for Mo, Ni, Sc, Ti, V, and W, respectively while an on-site exchange potential of J = 1.0 eV was used for all metals. These values were used in previous work.[9,26]

III. RESULTS

A. Structure Effects

Our previous study[26] revealed that dopant effects on physical and transport properties converge when the dopant-dopant interactions are minimized (*i.e.*, the dopants are effectively isolated). This coincides with a dopant concentration of 2.083 at.% ($M'M_{15}X_{32}$ where M' = dopant atom, M = Mo or W, X = S, Se, or Te). Therefore, this concentration was used for all systems in this study.

To explore a wide range of dopant effects on 2D MoX_2 and 2D WX_2 systems, six different transition-metal-site dopants were considered: Mo, Ni, Sc, Ti, V, and W. DFT-based structural relaxation calculations were employed to obtain ground state configurations for each doped system. In most cases, the structural distortion (or difference, Δ%, in lattice constant, bond lengths, and bond angles from that of the undoped) is minimal for 2.083 at.% dopant on 2D MoX_2 and 2D WX_2. For both structures, the Ni and Sc dopants cause the largest distortions (as much as 10.66% and 12.78% reductions in monolayer thickness (X-X bond lengths) for Ni in $MoTe_2$ and WTe_2, respectively). These two dopants have the lowest oxidation states of any metals investigated (+2 for Ni and +3 for Sc) and have respectively, the smallest and largest atomic radii. The tellurides consistently exhibit larger distortions than the other dichalcogenides. These distortions are most noticeable in the layer thickness, corresponding to X-X bond length (d_{X-X}), and X-M-X bond angle (θ). The least distorted structures are W-doped MoX_2 and Mo-doped WX_2. This is consistent with the fact that Mo and W have similar atomic radii (0.79 Å and 0.80 Å, respectively) and tend to have similar oxidation states (+6). Overall, the lattice constant, a_{uc}, experiences the least distortion of all measured geometries ($\leq |1.00\%|$ change). This implies that differences in layer thickness, rather than lattice mismatch, would be the primary cause for any potential dopant instability.

Another metric for comparing dopant stability is through the investigation of binding energy. A straightforward approach that offers quick comparison between a large number of doped structures follows that of our previous work[26], adapted from the work of Zhou et al.[16] This approach ignores strain energy and focuses on the energy involved with the dopant-chalcogen bonds. The equation takes the form:

$$E_b = E_{tot}^{doped} - E_{no\ M'}^{undoped} - E_{M'} \qquad (1)$$

where the total energy of the doped system, energy of the undoped system where the dopant atom (M') was removed, and the chemical potential of the dopant atom taken from its unary bulk structure are represented by E_{tot}^{doped}, $E_{no\ M'}^{undoped}$, and $E_{M'}$, respectively. By using this convention, negative binding energy values indicate that the doped system is a lower energy configuration than having the dopant atom separated from the system with no dopant at the M' site. Therefore, dopants with more negative (larger absolute value) binding energies are predicted to be more stable. The resulting binding energy values, given in Fig. 2, are depicted graphically versus the dopant atom for each 2D MX_2 system. The dopant atoms are listed in order of increasing atomic radius. Here we can see that the Ni dopant has the least negative binding energy among all systems (ranging from -1.60 eV in WTe_2 to -2.50 eV in MoS_2), indicating that it is the least stable dopant observed. This could be due to a couple of factors including: Ni has the lowest average oxidation state (+2) and it has the smallest atomic radius (0.62 Å). All other dopants exhibit large negative binding energies and are assumed to be stable. The Sc and Ti

978-1-5386-1202-6/19 $31.00 © 2019 IEEE

dopants have binding energies ranging from -4.58 eV (Sc) and -4.37 eV (Ti) in both telluride systems to -6.98 eV (Sc) and -7.44 eV (Ti) in MoS_2. This is in contrast to the fact that the Sc-doped systems have among the largest distortions, suggesting that atomic radius of the dopants plays a larger role in its stability than the oxidation state. All dopants in MoS_2, except Ni, have large negative binding energies (around -7 eV) suggesting high stability. However, the dopants show noticeably decreasing stability with increasing chalcogen atomic number. Also, the difference in binding energy between the MoX_2 and WX_2 systems is less than 1 eV. The largest difference is 0.86 eV, occurring between the V-doped MSe_2 systems.

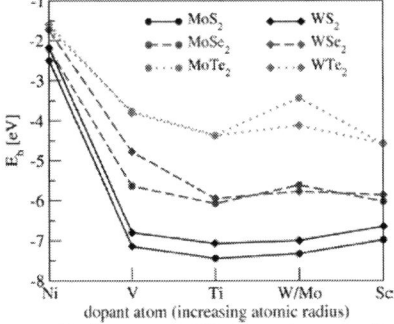

Fig. 2 Dopant atom binding energy (E_b) for each system listed in order of increasing dopant atomic radius. Solid black, dashed blue, and dotted red lines represent MS_2, MSe_2, and MTe_2 systems, respectively. Circle-lines depict MoX_2 structures while diamond-lines depict WX_2 structures. More-negative binding energies (greater absolute values) correspond to more-stable structures.

B. Electronic Structure

The electronic structures were calculated for each doped system and compared with that of the undoped systems. All undoped 2D MX_2 systems were found to be direct band gap semiconductors with gaps of 1.69, 1.84, 1.09, 1.84, 1.57, and 1.07 eV for MoS_2, $MoSe_2$, $MoTe_2$, WS_2, WSe_2, and WTe_2, respectively. The calculated band gap of MoS_2 is slightly smaller than the experimental value of 1.80 eV. The reason for slight underestimation is that we used computationally efficient GGA PBE approximation instead of computationally expensive hybrid method. This paper focuses on the dopant effect on electronic structure. For both MoX_2 and WX_2 systems, the band gaps generally decrease with increasing chalcogen atomic number. In most cases, the presence of 2.083 at.% transition metal dopant (*i.e.*, $M'M_{15}X_{32}$), reduces the band gaps or eliminates them, inducing metallic behavior in the compound. Fig. 3 plots the electronic density of states (DoS) for each Mo-, Ni-, Sc-, and W-doped system and compares them with the undoped DoS data. Specifically for the W-doped $MoSe_2$ (Fig. 3c), there is a reduction in the conduction band minimum (CBM) whereas all other Mo/W-doped systems have CBMs near that of their undoped structures. In fact, other than the $MoSe_2$ case, the W/Mo dopants have minimal impact on the electronic DoS of MoX_2/WX_2. This could be due to the similarity between the two metals (Mo and W have nearly the same atomic radius and oxidation states). Each of the Mo- and W-doped systems investigated here are direct band gap semiconductors with Fermi levels near the top of the valence

band, suggesting *p*-type behavior. Sc was the only dopant to produce a semiconducting material with a band gap different from the pure MX_2 compound (Sc-doped $MoSe_2$ has a small band gap of about 0.14 eV).

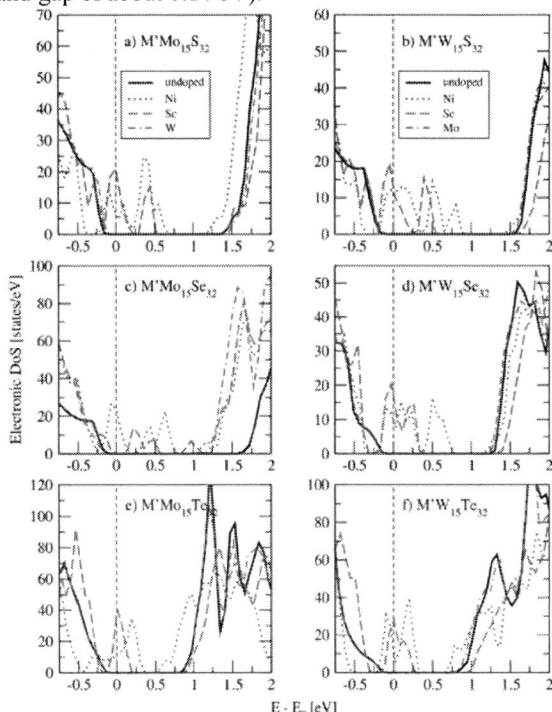

Fig. 3 Electronic density of states (DoS) for the $M'M_{15}X_{32}$ system (M' = Mo, Ni, Sc, and W; M = Mo or W; and X = S, Se, or Te). Column 1 (a, c, and e) contains the MoX_2 data while column 2 (b, d, and f) contains the WX_2 data. The rows depict sulfide (a and b), selenide (c and d), and telluride (e and f) systems, respectively. Undoped, Ni-, Sc-, and W/Mo-doped systems are represented by solid black, dotted blue, dashed red, and dash-dotted green lines, respectively. The x-axis is shifted to $E_F = 0$ eV for easy comparison.

C. Optical Properties

The electronic structures of 2D MX_2 are significantly modified after they are doped, indicating potential changes of the optical properties. In particular, the Sc- and Ni-doped systems have the strongest mid-gap states which may result in significant changes in their optical conductivities;[26] therefore, we have further calculated the effects of Sc and Ni dopants on the optical conductivities of MX_2. For comparison, we have also calculated the optical properties of W-doped MoX_2 and Mo-doped WX_2, of which the electronic structures are only slightly affected by the dopants. The real part of the optical conductivities of pure 2D MX_2 and Sc-, Ni-, W-, and Mo-doped 2D MX_2 systems are given in Fig. 4. The first absorption peaks in pure MoS_2 and WS_2 are found at about 2~3 eV, which agrees with previous theoretical studies[27,28] and experimental measurements.[29,30] For the systems doped with Sc or Ni, the amplitude of the first absorption peak becomes smaller while a new absorption peak emerges at low energies. This originates from the mid-gap states located just above the Fermi energy, consistent with the electronic DoS calculations. Our calculations suggest that the optical conductivities of the Sc-or Ni-doped 2D MX_2 are activated at much lower energies than those of the pure systems, resulting in enhanced photosensitivity and photoresponsivity. The positions of these new optical conductivity peaks induced by Sc or Ni doping are summarized in Table I. On the other hand, no obvious changes on the optical conductivities of W-doped MoX_2 and Mo-doped WX_2 are found in the low energy range, in agreement with the electronic DoS which exhibit no mid-gap states around the Fermi level. The photosensitivity and photoresponsivity of the W- and Mo- doped systems are not affected as much as those of the Sc- or Ni-doped systems.

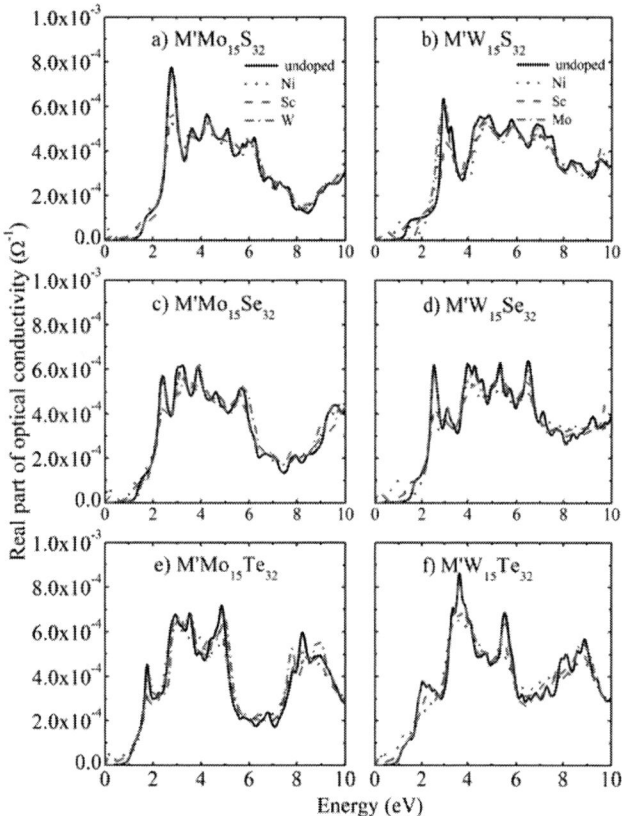

Fig. 4 . Real parts of the optical conductivities of pure and doped MX_2 with a dopant concentration of 2.083 at.%. The first column - plots (a), (c), and (e) - depict MoX_2 systems while the second column - plots (b), (d), and (f) - depict WX_2 systems. The rows represent sulfides (a and b), selenides (c and d), and tellurides (e and f), respectively. Data from undoped, Sc-, Ni-, and W-/Mo-doped systems are depicted with black, red, blue, and green lines, respectively.

Table I. Positions of the new absorption peaks (in eV) in the real part of optical conductivity induced by Sc and Ni dopants.

	Sc-doped	Ni-doped
MoS_2	0.36	0.17
$MoSe_2$	0.18	0.13
$MoTe_2$	0.15	0.22
WS_2	0.41	1.00
WSe_2	0.18	0.93
WTe_2	0.20	0.35

IV. CONCLUSION

DFT-based calculations were employed to investigate the structural and electrical properties of metal-site doped $M'M_{15}X_{32}$ (where M' = Mo, Ni, Sc, Ti, V, or W; M = Mo or W; and X = S, Se, or Te). The dopants mostly exhibit minimal structural distortion on 2D MX_2 with the exception of Ni and

Sc. Comparison of their geometry effects and binding energies suggests that their atomic radii may have a larger impact on dopant stability than their oxidation states. Binding energy results also indicate decreasing overall dopant stability with increasing chalcogen atomic number. Electronic structure calculations indicate that only W-doped MoX_2 and Mo-doped WX_2 are direct band gap semiconductors. All other doped systems (with the exception of Sc-doped $MoSe_2$ which has a small band gap of 0.14 eV) introduce mid-gap states on and above the Fermi level (which resides in the top of the valence band) implying a p-type metallic or conductive behavior. This goes against the doping propensity of MoS_2 which is experimentally determined to be n-type. For the semiconducting Mo- and W-doped systems, only the W-doped $MoSe_2$ system exhibited a significantly reduced band gap compared to the undoped. The photosensitivity and photoresponsivity of 2D MX_2 are enhanced by doping with Sc or Ni due to the strong mid-gap states induced by the dopants. This work explores various dopants and their effects on 2D MX_2 structures which is useful for guiding development of these materials for electronic and device applications.

ACKNOWLEDGMENT

Computational resources for this work were provided by Boise State University's R1 cluster and Idaho National Laboratory's high performance computing (HPC) facility. The authors would like to thank the Research Experience for Undergraduates (REU) program at Boise State University as well as Ken Blair, Tyler Bevan, and Cormac Garvey for technical support. SL and YC are grateful for the research computing facilities offered by ITS, HKU.

REFERENCES

(1) Coleman, J. N.; Lotya, M.; O'Neill, A.; Bergin, S. D.; King, P. J.; Khan, U.; Young, K.; Gaucher, A.; De, S.; Smith, R. J.; Shvets, I. V.; Arora, S. K.; Stanton, G.; Kim, H. Y.; Lee, K.; Kim, G. T.; Duesberg, G. S.; Hallam, T.; Boland, J. J.; Wang, J. J.; Donegan, J. F.; Grunlan, J. C.; Moriarty, G.; Shmeliov, A.; Nicholls, R. J.; Perkins, J. M.; Grieveson, E. M.; Theuwissen, K.; McComb, D. W.; Nellist, P. D.; Nicolosi, V., Two-Dimensional Nanosheets Produced by Liquid Exfoliation of Layered Materials, *Science* **2011**, *331*, 568.

(2) Gong, C.; Colombo, L.; Wallace, R. M.; Cho, K., The Unusual Mechanism of Partial Fermi Level Pinning at Metal-MoS2 Interfaces, *Nano Lett* **2014**, *14*, 1714.

(3) Butler, S. Z.; Hollen, S. M.; Cao, L. Y.; Cui, Y.; Gupta, J. A.; Gutierrez, H. R.; Heinz, T. F.; Hong, S. S.; Huang, J. X.; Ismach, A. F.; Johnston-Halperin, E.; Kuno, M.; Plashnitsa, V. V.; Robinson, R. D.; Ruoff, R. S.; Salahuddin, S.; Shan, J.; Shi, L.; Spencer, M. G.; Terrones, M.; Windl, W.; Goldberger, J. E., Progress, Challenges, and Opportunities in Two-Dimensional Materials Beyond Graphene, *Acs Nano* **2013**, *7*, 2898.

(4) Wang, Q. H.; Kalantar-Zadeh, K.; Kis, A.; Coleman, J. N.; Strano, M. S., Electronics and optoelectronics of two-dimensional transition metal dichalcogenides, *Nat Nanotechnol* **2012**, *7*, 699.

(5) Huang, W.; Luo, X.; Gan, C. K.; Quek, S. Y.; Liang, G. C., Theoretical study of thermoelectric properties of few-layer MoS2 and WSe2, *Physical Chemistry Chemical Physics* **2014**, *16*, 10866.

(6) Huang, W.; Da, H. X.; Liang, G. C., Thermoelectric performance of MX2 (M5M(o),W; X = S, Se) monolayers, *J Appl Phys* **2013**, *113*, 104304.

(7) Tributsch, H., Solar Energy-Assisted Electrochemical Splitting of Water, *Z Naturforsch A* **1977**, *32*, 972.

(8) Lin, Z.; Carvalho, B. R.; Kahn, E.; Lv, R.; Rao, R.; Terrones, H.; Pimenta, A. M.; Terrones, M., Defect Engineering of Two-dimensional Transition Metal Dichalcogenides, *2D Mater.* **2016**, *3*, 022002.

(9) Williamson, I.; Correa Hernandez, A.; Wong-Ng, W.; Li, L., High-Throughput Computational Screening of Electrical and Phonon

Properties of Two-Dimensional Transition Metal Dichalcogenides, *JOM* **2016**, *68*, 2666.

(10) Shanavas, K. V.; Satpathy, S., Effective Tight-Binding Model for MX$_2$ under Electric and Magnetic Fields, *Phys. Rev. B* **2015**, *91*, 235145.

(11) Onofrio, N.; Guzman, D.; Strachan A., Novel Doping Alternatives for Single-Layer Transition Metal Dichalcogoneides, *J. Appl. Phys.* **2017**, *122*, 185102.

(12) Nourbakhsh, A.; Zubair, A.; Sajjad, R. N.; Tavakkoli K. G., A.; Chen, W.; Fang, S.; Ling, X.; Kong, J.; Dresselhaus, M. S.; Kaxiras, E.; Berggren, K. K.; Antoniadis, D.; Palacios, T., MoS$_2$ Field-Effect Transistor with Sub-10 nm Channel Length, *Nano Lett.* **2016**, *16*, 7798.

(13) Kuc, A.; Zibouche, N.; Heine, T., Influence of quantum confinement on the electronic structure of the transition metal sulfide TS2, *Phys Rev B* **2011**, *83*.

(14) Splendiani, A.; Sun, L.; Zhang, Y. B.; Li, T. S.; Kim, J.; Chim, C. Y.; Galli, G.; Wang, F., Emerging Photoluminescence in Monolayer MoS2, *Nano Lett* **2010**, *10*, 1271.

(15) Mak, K. F.; Lee, C.; Hone, J.; Shan, J.; Heinz, T. F., Atomically Thin MoS2: A New Direct-Gap Semiconductor, *Phys Rev Lett* **2010**, *105*.

(16) Zhou, Y. G.; Su, Q. L.; Wang, Z. G.; Deng, H. Q.; Zu, X. T., Controlling magnetism of MoS2 sheets by embedding transition-metal atoms and applying strain, *Physical Chemistry Chemical Physics* **2013**, *15*, 18464.

(17) Cheng, Y. C.; Zhu, Z. Y.; Mi, W. B.; Guo, Z. B.; Schwingenschlogl, U., Prediction of two-dimensional diluted magnetic semiconductors: Doped monolayer MoS2 systems, *Phys Rev B* **2013**, *87*, 100401.

(18) Chen, Y. F.; Xi, J. Y.; Dumcenco, D. O.; Liu, Z.; Suenaga, K.; Wang, D.; Shuai, Z. G.; Huang, Y. S.; Xie, L. M., Tunable Band Gap Photoluminescence from Atomically Thin Transition-Metal Dichalcogenide Alloys, *Acs Nano* **2013**, *7*, 4610.

(19) Suh, J.; Park, T. E.; Lin, D. Y.; Fu, D. Y.; Park, J.; Jung, H. J.; Chen, Y. B.; Ko, C.; Jang, C.; Sun, Y. H.; Sinclair, R.; Chang, J.; Tongay, S.; Wu, J. Q., Doping against the Native Propensity of MoS2: Degenerate Hole Doping by Cation Substitution, *Nano Lett* **2014**, *14*, 6976.

(20) Hsu, W. K.; Zhu, Y. Q.; Yao, N.; Firth, S.; Clark, R. J. H.; Kroto, H. W.; Walton, D. R. M., Titanium-doped molybdenum disulfide nanostructures, *Advanced Functional Materials* **2001**, *11*, 69.

(21) Sun, Q. C.; Yadgarov, L.; Rosentsveig, R.; Seifert, G.; Tenne, R.; Musfeldt, J. L., Observation of a Burstein-Moss Shift in Rhenium-Doped MoS2 Nanoparticles, *Acs Nano* **2013**, *7*, 3506.

(22) Graedel, T. E.; Harper, E. M.; Nassar, N. T.; Nuss, P.; Reck, B. K., Criticality of metals and metalloids, *P Natl Acad Sci USA* **2015**, *112*, 4257.

(23) Kresse, G.; Furthmuller, J., Efficient iterative schemes for ab initio total-energy calculations using a plane-wave basis set, *Phys Rev B* **1996**, *54*, 11169.

(24) Perdew, J. P.; Burke, K.; Ernzerhof, M., Generalized gradient approximation made simple, *Phys Rev Lett* **1996**, *77*, 3865.

(25) Liechtenstein, A. I.; Anisimov, V. I.; Zaanen, J., Density-functional theory and strong interactions: Orbital ordering in Mott-Hubbard insulators, *Phys Rev B* **1995**, *52*, R5467.

(26) Williamson, I.; Li, S.; Correa Hernandez, A.; Lawson, M.; Chen, Y.; Li, L., Structural, electrical, phonon, and optical properties of Ti- and V-doped two-dimensional MoS2, *Chemical Physics Letters* **2017**, *674*, 157.

(27) Kumar, A.; Kumar, J.; Ahluwalia, P. K., Electronic Structure and Optical Conductivity of Two Dimensional (2D) MoS2: Pseudopotential DFT Versus Full Potential Calculations, *AIP Conference Proceedings* **2012**, *1447*, 1269.

(28) Carvalho, A.; Ribeiro, R. M.; Neto, A. H. C., Band nesting and the optical response of two-dimensional semiconducting transition metal dichalcogenides, *Phys Rev B* **2013**, *88*.

(29) Nayak, P. K.; Yeh, C. H.; Chen, Y. C.; Chiu, P. W., Layer-Dependent Optical Conductivity in Atomic Thin WS2 by Reflection Contrast Spectroscopy, *Acs Appl Mater Inter* **2014**, *6*, 16020.

(30) Milosevic, I.; Nikolic, B.; Dobardzic, E.; Damjanovic, M.; Popov, I.; Seifert, G., Electronic properties and optical spectra of MoS(2) and WS(2) nanotubes, *Phys Rev B* **2007**, *76*.

Izaak Williamson received his PhD in materials science and engineering in 2017 from Boise State University in Boise, Idaho. He completed a post-doctoral appointment as a Technology Development Materials Modeling Engineer at Micron Technology, Inc.

(Boise, Idaho) in 2018. He now works as a metrology process development engineer for Micron where he supports OCD and scatterometry applications.

Dr. Williamson has contributed to over 14 publications including 4 authored scientific journal articles and 2 textbook chapters. He has also given 10 technical conference presentations.

Matthew Lawson was born in Johannesburg, South Africa in 1994. He received his B.S. degree in materials science and engineering from Boise State University in Boise, Idaho, in 2017, and is currently pursuing his Ph.D. degree in materials science and engineering at Boise State University.

From beginning his Ph.D. in 2017 he works with low dimensionality materials, primarily TMDs. His works pertains not only to first-principles calculations property screening but also atomic layer deposition characterization and understanding initial growth mechanics.

Shasha Li received her M.S. degree in Condensed Matter Physics from Henan Normal university in 2015, and is currently pursuing her PhD degree in Mechanical Engineering at The University of Hong Kong. During her PhD study, she investigates the lattice dynamics, and the anharmonic phonon-phonon coupling of thermoelectric materials using first-principles calculations.

Yue Chen is an assistant professor at the University of Hong Kong, Department of Mechanical Engineering. His research interests focus on the materials physics for electrical and thermal transports, such as electronic structures and lattice dynamics. His interests stem from the studies of materials science at Oxford University and Beihang University. He was a postdoctoral fellow at Columbia University in the City of New York and Institute of Metal research, Chinese Academy of Sciences before joining HKU.

Lan Li received her PhD in Materials Engineering at the University of Cambridge in the UK. Dr. Li worked as a post-doc associate and researcher at the National Institute of Standards and Technology (NIST), Kent State University, and University of Florida. Since Nov 2012, she has been an assistant professor at the Micron School of Materials Science and Engineering, Boise State University.

Dr. Li's awards and honors include NIST's American Recovery and Reinvestment Act Program Fellowship, TMS Young Leader Professional Development Award, William Mong Visiting Research Fellowship in Engineering at the University of Hong Kong

978-1-5386-1202-6/19 $31.00 © 2019 IEEE

Contributed Paper

Structural and Magnetic Properties of CoPd Alloys for Non-Volatile Memory Applications

Joseph B. Abugri, Billy D. Clark, Sujan Budhathoki, Pieter B. Visscher, *Senior Member, IEEE*, Adam Hauser, *Member, IEEE* and Subhadra Gupta., *Senior Member, IEEE*

Abstract—We have intensively studied perpendicular magnetic anisotropy (PMA) CoPd alloys, both structurally and magnetically, as a simple means of pinning MgO-based perpendicular magnetic tunnel junctions (pMTJs) for spin-transfer torque magnetic tunnel junction (STT-MRAM) applications. A compositional study of the Co_xPd_{100-x} alloys at 50 nm thickness showed that the maximum coercivity and anisotropy was found for $Co_{25}Pd_{75}$. Perpendicular magnetic tunnel junction stacks were deposited using different compositions of CoPd. Current-in-plane tunneling measurements indicated that the TMR values roughly correlated with the coercivity and anisotropy of the single layers. A thickness study indicated that the alloy was fully perpendicular for thicknesses as low as 20 nm. Various seed layers were employed to optimize the coercivity of the $Co_{25}Pd_{75}$ layer. Magnetometry, X-ray diffraction (XRD) and scanning electron microscopy (SEM) studies were carried out to relate the magnetic and structural properties of these layers.

Index Terms—perpendicular magnetic tunnel junctions, CoPd alloys, coercivity, anisotropy, seed layers, XRD, SEM.

I. INTRODUCTION

IN recent years there has been a great deal of excitement and research regarding spin-transfer torque magnetic random memory (STT-MRAM) as a potential future non-volatile memory. We have focused on perpendicular magnetic anisotropy (PMA) materials for fabricating fully perpendicular magnetic tunnel junctions (pMTJs) that are a key component of STT-MRAM.

In particular, we have focused on CoFeB-MgO based

This work was supported in part by the Center for Materials for Information Technology (MINT Center), Department of Physics and Astronomy, Department of Electrical and Computer Engineering, and Department of Metallurgical and Materials Engineering, all at The University of Alabama.

Joseph A. Abugri is with the Department of Electrical and Computer Engineering, The University of Alabama, Tuscaloosa, AL 35487, USA (email: jbabugri@crimson.ua.edu)

Billy D. Clark was with the Department of Electrical and Computer Engineering, The University of Alabama, Tuscaloosa, AL 35487, USA. He is currently with Intel Corporation, Hillsboro, OR   , USA (email: billy.clark@intel.com).

Pieter B. Visscher, Sujan Budhathoki and Adam Hauser are with the Department of Physics and Astronomy at The University of Alabama, Tuscaloosa, AL 35487, USA. (emails: visscher@bama.ua.edu, sbudhathoki@crimson.ua.edu, ahauser@ua.edu)

Subhadra Gupta, the corresponding author, is with the Department of Metallurgical and Materials Engineering, The University of Alabama, Tuscaloosa, AL 35487, USA (email: sgupta@eng.ua.edu).

pMTJs, with perpendicular Co/Pd multilayer synthetic anitiferromagnet (SyAF) pinning of the reference layer of CoFeB.[1-3] This is now a commonly used scheme for high quality pMTJ's in industry.[4] However, the deposition of Co/Pd multilayers of precise thickness and interfacial quality can be a challenge, and is time consuming. With that in mind, we have investigated single CoPd alloys as a function of composition, thickness, deposition and annealing temperatures, and seed layers. The details of the composition study are elaborated in Reference 5, but we will review some of the essential results in this paper.

II. EXPERIMENT

An SFI Shamrock planetary, sputter-up, load-locked, fully automated system was utilized to deposit the CoPd alloy films as well as the pMTJ stacks onto 75 mm silicon substrates. The sputtering system was equipped with six targets and an infrared lamp heater, and typically achieved a base pressure of less than 4×10^{-6} Pa. All the films were deposited onto thermally oxidized silicon substrates by dc magnetron sputtering, except for MgO, which was rf sputtered from a compound target. A sputter pressure of 0.26 Pa was used for all the depositions except for the CoPd alloys. The Co_xPd_{100-x} alloys were co-deposited from elemental targets at an argon pressure of 0.8 Pa. Deposition rates for the individual targets were derived from thickness estimations from x-ray reflectivity (XRR) measurements. For the co-deposited CoPd alloys, the target powers used were estimated from these measured deposition rates, where the Co/Pd ratio was converted from a volume percentage to an atomic percentage, and finally confirmed by energy dispersive or wave dispersive spectroscopy (EDS or WDS). The film stacks were post-deposition annealed in-situ at 400 ^{0}C for 5 minutes, using the infrared lamp array.

Magnetic characterization was carried out on a Princeton Scientific alternating gradient magnetometer (AGM). In-plane and out-of plane M-H loops, as well as first order reversal curves (FORC) were measured.[6] A new open-source software, FORC+, was developed to plot both the irreversible and reversible parts of the FORC distribution.[7] Structural characterization was carried out using a Philips X'Pert x-ray diffraction (XRD) with Cu Kα radiation, and scanning electron microscopy (SEM) using a JEOL 7000 field emission SEM. Separate manuscripts on a transmission electron microscopy (TEM) study to elucidate the effect of seed layers[8] and a magnetic force microscopy (MFM) study to investigate the switching behavior of the CoPd alloys are in preparation.[9]

III. RESULTS AND DISCUSSION

The initial studies of CoPd as a function of composition were carried out with 50 nm of CoPd sandwiched between a seed layer of 13 nm MgO and a capping layer of 3 nm MgO. The highest perpendicular magnetic anisotropy of about 7×10^5 ergs/cm^3 was found for a nominal composition of $Co_{25}Pd_{75}$. The coercivity was seen to peak at about 30 at% Co.

Bottom-pinned perpendicular magnetic tunnel junction stacks (pMTJs) were deposited with various Co concentrations. The deposited stacks were $Si/SiO_2/MgO$ (13)/Co_xPd_{100-x} (50)/ Ta (0.3) /CoFeB (1)/MgO (1.6)/CoFeB (1)/Ta (5)/Ru (10), with the numbers in parentheses being the layer thicknesses in nm. These stacks were in-situ lamp annealed after deposition. The unpatterned pMTJ stacks were measured by current-in-plane tunneling at NIST Gaithersburg, and found to have a room temperature tunneling magnetoresistance (TMR) value of 50% for the $Co_{25}Pd_{75}$ alloy-pinned MTJ.

In the present work, we tried to improve these properties (obtained for 50 nm thickness and MgO seed) by varying the thickness and the seed layer. The MgO barrier layer would also need to be improved to obtain higher TMR, but our focus here was on the CoPd alloy layer. We performed a study of PMA vs. thickness of the optimized $Co_{25}Pd_{75}$ alloy in order to bring the thickness down to the thinnest value feasible that would still have sufficiently high anisotropy energy, K_u, and squareness, S^* (M_r/M_s).

A. Thickness study

A series of samples with the thickness of the $Co_{25}Pd_{75}$ layer varied from x=10 to 50 nm in 10 nm increments was sputtered. The stack configuration was $Si/SiO_2/MgO$ (13)/

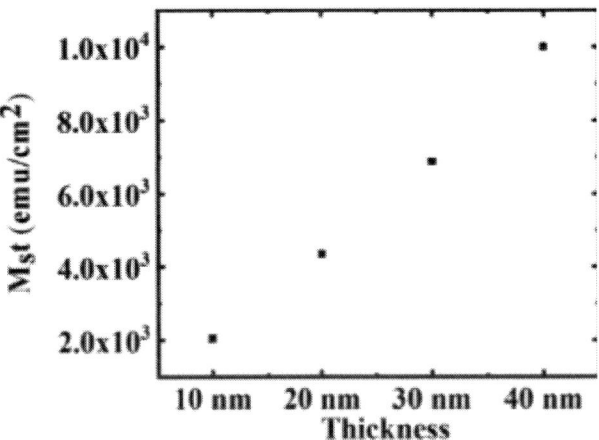

Fig. 2. $M_s t$ vs. t is a straight line through the origin, as expected.

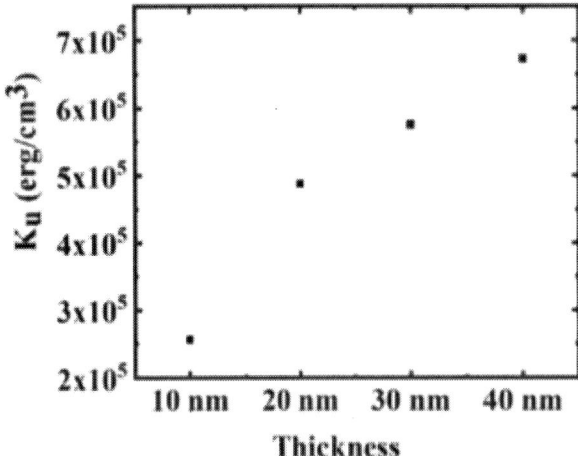

Fig. 3. K_u vs. t indicates that the perpendicular anisotropy increases with thickness, but is still reasonable for 20 nm.

$Co_{25}Pd_{75}$ (x)/Ta (5), with the values in parentheses being nm. The out-of-plane M-H loops are shown in Figure 1, indicating that the alloy is fully perpendicular down to 20 nm. At 10 nm, the M-H loop still shows easy-axis behavior, but the remanence is too low to be useful as a perpendicular pinning layer for CoFeB.

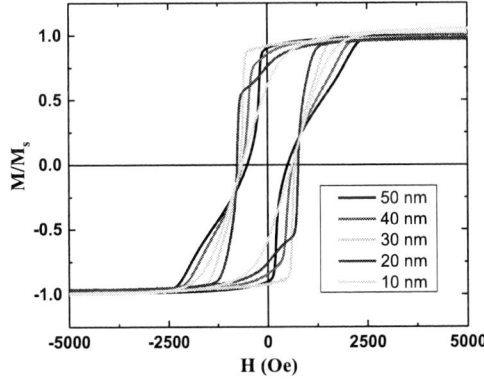

Fig. 1. Out-of-plane hysteresis loops of $Si/SiO_2/Co_{25}Pd_{75}/Ta$ with varying thicknesses of the CoPd alloy.

Figures 2, 3 and 4 show respectively, $M_s t$ vs. t, where M_s is the saturation magnetization and t is the CoPd film thickness, K_u vs. t, where $K_u = (M_s H_k)/2$ is the anisotropy energy density
and H_k is the anisotropy field, and S* vs. t, where S* = M_r/M_s is the squareness of the loop.

Scanning electron micrographs of the 10, 20, 30, and 40 nm

Fig. 5. SEM micrographs of a) 10nm, b) 20 nm, c) 30 nm, d) 40 nm $Co_{25}Pd_{75}$ films.

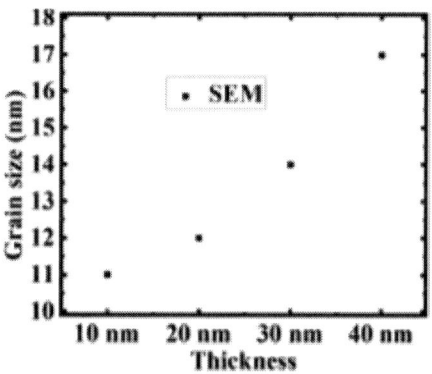

Fig. 6. Grain size estimation from scanning electron microscopy

film stacks are shown in Figure 5. Grain size is estimated from these micrographs using ImageJ software, and plotted in Fig. 6. Since the x-ray diffraction peaks of the thin CoPd films are difficult to distinguish, the SEM micrographs yield a clearer idea of the grain size than the Debye-Scherrer analysis[10] of the x-ray spectra. The coercivity of the films decrease as the film thickness increases from 20 to 50 nm. This agrees with results found in Reference 11, where the coercivity decreases as the crystallite size increases. This trend is attributed to the larger number of grain boundaries for thinner films, and the coercivity may be related to domain wall motion across grain boundaries. The 10 nm film deviates from this trend, and this

may be caused the tendency for it to go in-plane at such low thicknesses.

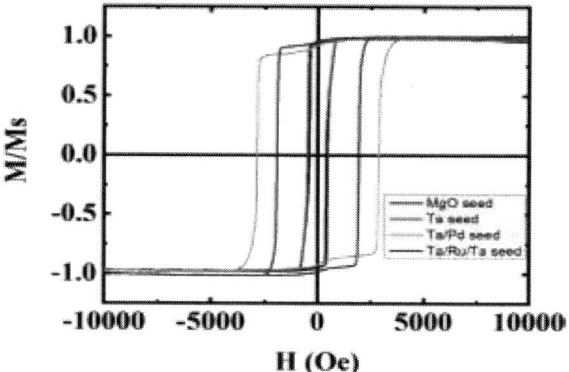

Fig. 7. Out-of-plane M-H loops with different seed layers

B. Seed layer study

From the thickness study detailed above, it was determined that we could reduce the CoPd thickness to 20 nm, and still have enough perpendicular anisotropy to pull 1 nm of CoFeB

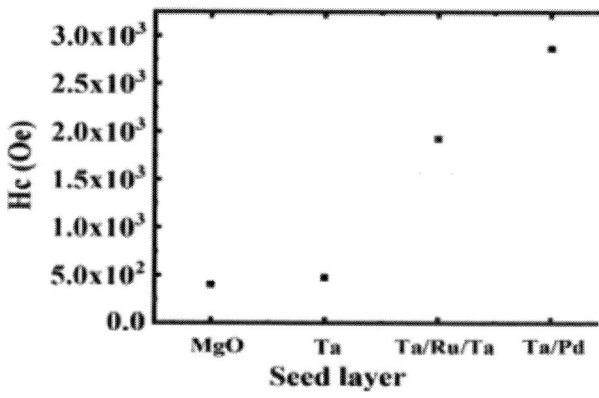

Fig. 8. Coercivity for various seed layers

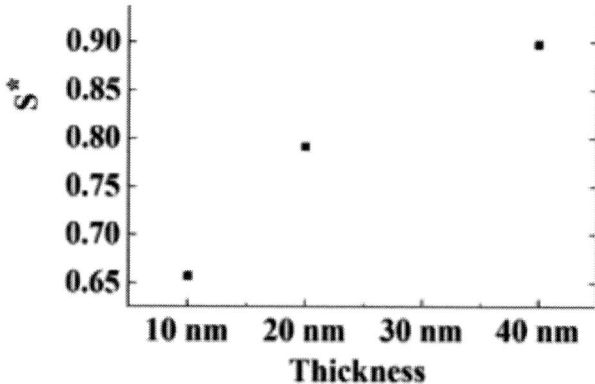

Fig. 4. The squareness reaches a maximum at 30 nm, then decreases.

out of plane.

We then embarked on a study of seed layers for further optimization of these 20 nm $Co_{25}Pd_{75}$ films. The seed layers investigated were, MgO (5 nm), Ta (5 nm), Ta (5)/Ru (5/Ta (5) nm and Ta (5)/Pd (5) nm. The stacks deposited were Si/SiO_2/seed layer(s)(x)/ CoPd (20)/Ta (5 nm). To our surprise, the coercivity of the samples varied greatly, as shown in Figures 7 and 8, respectively, from about 400 Oe for the MgO seed to nearly 3 kOe for the Ta/Pd seed.

AFM studies showed that the surface roughness of all the films was less than 0.5 nm, so that did not explain the change in coercivity.

X-ray diffraction was carried out on these samples, as shown in Figure 9. The presence and position of CoPd (111) ($2\theta = 40.7^{\circ}$) peaks[12, 13] indicate that all seed layers yield fcc (111) texture, and are not epitaxially straining the CoPd layer. All samples show CoPd (200), CoPd (111), and Ta (202) peaks at 40.7, 43.2, and 37.1 degrees, respectively.

We note that the films with the strongest CoPd (111) intensity, and ostensibly the best crystal quality, also have higher magnetic coercivity. While not definitive proof, the

results suggest that the higher crystallinity films should possess higher coercivity. In the case of the heterostructure with Ta/Pd seed layer, one may be able to attribute the coercivity increase to higher atomic content of Pd in the stack, as previously observed. However, the heterostructure with Ta/Ru/Ta seed layer also has strongly enhanced coercivity, suggesting that interfacial quality and crystallinity could play a major role as well.

Scanning electron micrographs were taken of the 20 nm CoPd on the various seed layers, as shown in Figure 10. Grain sizes for the samples with the various seed layers are shown in Figure 11, as estimated by ImageJ software. The samples with the highest coercivity appear to also have the finest grains, as discussed above.

Grain sizes were also plotted in Figure 11 from the XRD spectra using the Debye-Scherrer analysis. There is much less of a change in grain size, and the grains are much larger. Since

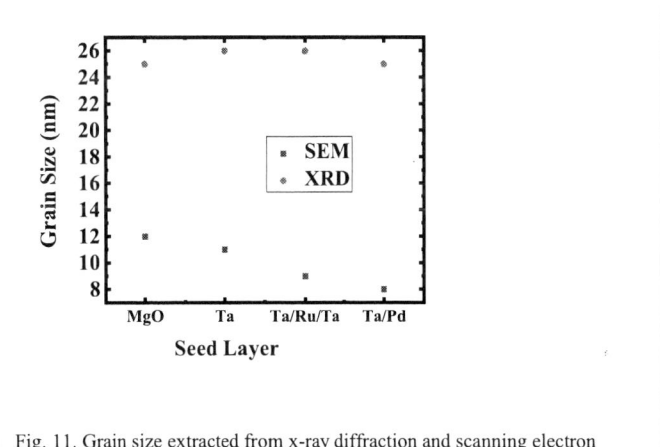

Fig. 11. Grain size extracted from x-ray diffraction and scanning electron

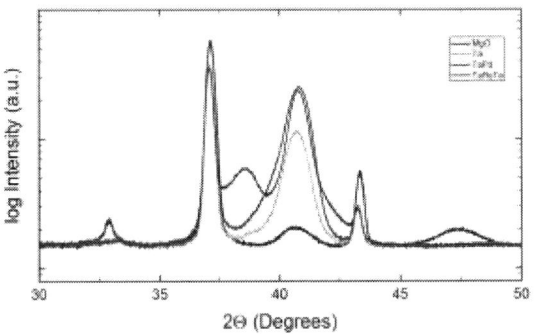

Fig. 9. X-ray diffraction spectra of CoPd films with various seed layers.

this analysis is representative of grains in the z direction, whereas the data from the SEM are representative of grains in the x-y direction, the difference is understandable.

CONCLUSIONS

We have conducted a thickness optimization as well as a seed layer optimization for $Co_{25}Pd_{75}$ alloys. The coercivity of the CoPd films was found to be strongly dependent on the crystal structure of the CoPd grown on different seed layers, with the highest coercivity obtained for the finest grain size. Cross-sectional transmission electron microscopy will be carried out in future to further elucidate the magnetic behavior of these films.

IV. ACKNOWLEDGEMENTS

The authors acknowledge the MINT Center, the Department of Electrical and Computer Engineering, the Department of Physics and Astronomy, and the Department of Metallurgical and Materials Engineering for support.

REFERENCES
1. D.C. Worledge, G. Hu, D. W. Abraham, J. Z. Sun, P. I. Trouilloud, J. Nowak, S. Brown, M. C. Gaidis, E. J.

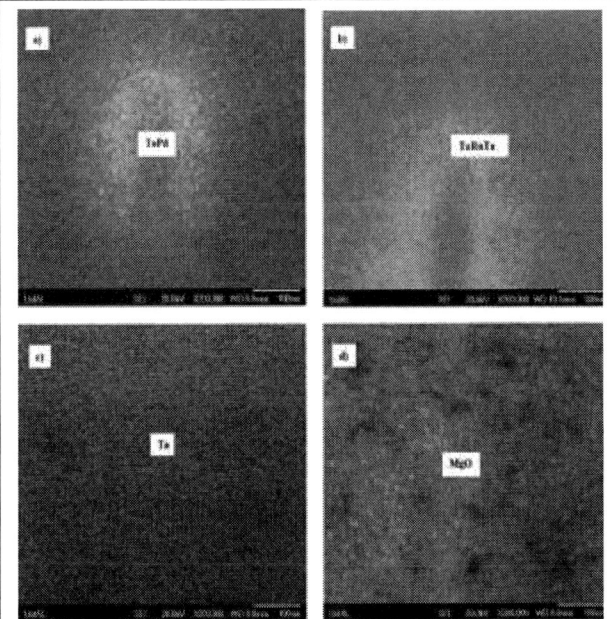

Fig. 10. Scanning electron micrographs of CoPd grown on a) Ta/Pd, b) Ta/Ru/Ta, c) Ta, and d) Pd seed layers.

O'Sullivan, and R. P. Robertazzi, Appl. Phys. Lett. **98** (2011),022501-1.

2. A. Natarajarathinam, R. Zhu, P.B. Visscher, and S. Gupta, J. Appl. Phys. **111** (2012), 07C918-1.

3. A. Natarajarathinam, B.D. Clark, A. Singh and S. Gupta, J. Phys. D:Appl. Phys. **46** (2013), 095002.

4. M. Pakala et al., 63rd AVS Symposium, Tampa, FL (2017).

5. B.D. Clark, A. Natarajarathinam, Z. R. Tadisina, P. J. Chen, R. D. Shull, and S. Gupta, J. Magn. Magn. Mat'ls. **436** (2017), 113.

6. J. B. Abugri, P. B. Visscher, S. Gupta, P. J. Chen and R. D. Shull, J. Appl. Phys. **124** (2018), 043901 (2018).

7. P. B. Visscher, FORC+ Software, http://MagVis.org.

8. J. B. Abugri, B.D. Clark, Philomela Komninou, P.B. Visscher and S. Gupta, in preparation (2019).

9. J. B. Abugri, B.D. Clark, P.B. Visscher, J. Gong, E. Singleton, and S. Gupta, in preparation (2019).

10. B.D. Cullity, "Elements of X-Ray Diffraction", 2nd ed. (1978).

11. A. Bourezg and A. Kharmouche, Vacuum **155** (2018), 612.

12. International Center for Diffraction Data (ICDD) database.

13. C. Morgan, K. Schmalbuch, F. Garcia-Sanchez, C. M. Schneider and Carola Meyer, J. Magn. Magn. Mat'ls., **325** (2013), 112.

Contributed Paper

3D Simulation Technique To Predict Failure of Photo Marks Interaction

Radhakrishna Kotti, IEEE Member, Micron Technology, Boise, Idaho

Abstract—**This paper presents a 3D simulation technique to predict the interaction failure of scribe photo marks. Device scaling for increased speed and density, coupled with advanced memory architectures is increasing the fabrication complexity. These process changes are accompanied by additional masking levels with multiple layers to align. Each new masking level requires photo marks such as registration, alignment and metrology for their processing. The process is first optimized for the die layout leaving the construction of photo marks vulnerable due to their unique patterning requirements. This may cause some unintended process interactions creating registration or alignment failures. So, a validation technique guided by modeling mask and process interaction is proposed to avoid unintended structural results. The method proposed is implemented and verified using a 3D simulator on alignment and registration marks. The results obtained from the proposed method clearly show that any new unintended process interaction can be identified at the pre-silicon stage and rectified before reticle manufacture improving the learning time and cost.**

Index Terms—**Photo lithography, 3D simulation, Scribe Marks, Mask Interaction, Design Rule Checks, Registration marks, Alignment marks, Scatterometry marks.**

I. INTRODUCTION

In order to optimize the use of rapid increase in information provided by the distributed computer networks, there is a continuous effort to improve the computational capability of these systems. But there has been a continuous gap in the performance of processing, memory and storage systems [1]. Today the gap in improvement between processing and memory is more than 50% and this is the bottleneck to improve the overall performance of the system. This gap further increases when the memory has an integrated storage unit, as shown in Fig.1.

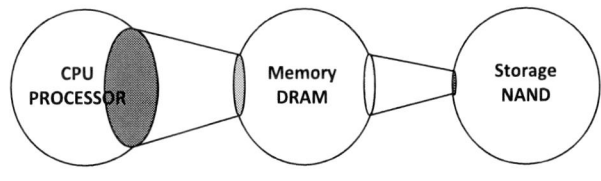

Fig. 1.   Performance bottlenecks between processor, memory and storage

These bottlenecks can be improved by advancements in memory and storage technologies. This demand in performance with higher density and lower cost per bit leads to their down scaling [2]. This scaling and shift in architecture is often accomplished by moving form 2D-planar to 3D stack. These advancements sometimes come at the cost of larger die size reducing the available scribe space [3,4]. Scribe is the space between adjoining die which provides the real estate for test structures, process control monitors and photo marks. The scribe structures pattern density differs from the die and creates process variations due to the process being optimized for the die design [5]. The scribe is diced during packaging and may not add real value to the final product. So, this is kept as small as possible to increase the die per wafer.

This increase in process complexity requires additional mask levels or process steps [6]. Each new masking level requires photo marks such as registration, alignment and metrology [7,8]. In order to accommodate the real estate for these new photo marks we may need to reuse the same scribe location at different process levels making sure there will not be any un-intended mask interactions causing yield issues or process defects. It is really hard to foresee all possible mask interactions due to additional levels and predict the failure of photo marks on the reused scribe space. Generally, these defects or failures are seen in the post-silicon processing increasing the cost and cycle time. So, in this paper a pre-silicon verification technique using new 3D simulation guided by mask interaction compare is proposed to overcome this limitation. This technique identifies the areas of concern in the layout and simulates the process outcome to help predict the defects prior to silicon reducing the process issues, cost and thereby the cycle time.

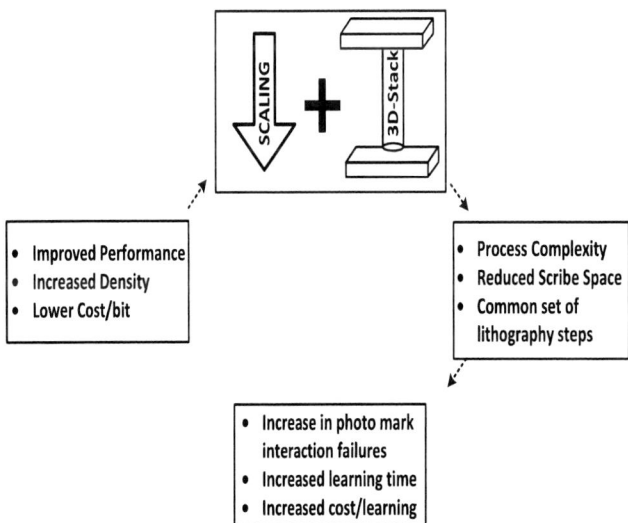

Fig. 2.   Advancements leading to increase in photo mark interaction failures

II. Proposed Method

The proposed method is a two step approach, first we need to detect the areas of concern and then run 3D process simulation at those locations to validate the issue. The area of concern is identified by running a mask interaction analysis in the layout. Those highlighted areas are analyzed in a 3D simulator which gives step by step process implementation at that location.

A. Mask interaction analysis

This analysis can be done in any layout viewer which can load mask design rule checks (DRCs). We need to set up mask DRC that identify and highlights any unintended mask interactions. In order to demonstrate this interaction analysis, we shall consider two interconnect levels that has via's (V), metal's (M) and etch (E) process steps. The intended steps of our process are V1->M1->E1 for first interconnect level and V2->M2->E2 for the second interconnect level. The intent of etch layer E is to chop only its corresponding Metal layer M and every via V has to be covered with top and bottom M layer. In order to make sure there are no unintended mask interactions we have the following mask DRC's:

1) V1 NOT INTERACT M1 OR M2: This DRC helps us identify if there are any areas in the layout where V1 has no interaction with M1 or M2. This helps us make sure there are no un-landed or uncovered via's. This is not expected in the die to which the process is optimized.

2) E1 INTERACT V1 NOT M1: This DRC helps us identify if there are any areas in the layout where E1 has interactions with V1 but not with M1. This helps us make sure that we examine the etch interaction with the metal dielectric and via surface. This is not expected in the die to which the process is optimized.

3) V2 NOT INTERACT M2: This DRC helps us identify if there are any areas in the layout where V2 has no interaction with M2.

4) E2 INTERACT V2 NOT M2: This DRC serves the similar purpose of E1 INTERACT V1 NOT M1, but for 2nd interconnect levels.

5) E2 INTERACT M1 NOT V2 OR M2: This DRC helps us identify if there are any areas in the layout where E2 has interactions with M1 but not with V2 or M2. This helps us make sure that we examine the etch interaction with the 2nd interconnect dielectric and M1. This is not expected in the die to which the process is optimized.

6) E2 INTERACT V1 NOT M1 OR V2 OR M2: This DRC helps us identify if there are any areas in the layout where E2 has interactions with V1 but not with M1 or V2 or M2. This helps us make sure that we examine the etch interaction with the 2nd interconnect dielectric and V1. This is a worst case

scenario and is not expected in the die to which the process is optimized.

All these interaction as shown in Fig.3 helps us identify the worst case mask interactions possible in the layout. 3D simulations are run on these highlighted interactions to validate the process steps to see if any of the highlighted areas will be a potential concern on wafer.

Fig. 3. Results for unintended Mask interactions highlighted by Mask DRC's

B. 3D simulation of unintended Mask interaction areas

The simulations for the areas of concern are ran on a 3D simulator to validate the concerns raised. This simulator gives us a step by step process simulation and helps us identify if any of the scribe structures like registration, alignment or metrology marks will be a source for defects or yield issues.

III. Results

As shown in Fig.3 there are 2 areas of concern the unintended mask interactions highlighted to be analyzed in 3D simulator. The first one is a alignment mark for V1 mask level which had a interaction with Mask E1 without any Mask M1 interaction in between. This 3D simulation gives us a step by step analysis to determine whether the V1 alignment mark getting etched at Mask M1 will be concern post E1. From the Fig.4 you can see that the simulation has identified surface area roughness at the center of the marks post E1 etch.

The second one is a M2-V2 registration mark which have a interaction with Mask E2. On the die, V2 is expected to be covered by M2 and later M2 is expected to be etched at E2. But here the V2 will be exposed to E2 etch directly without any M2 in between. This step by step analysis helps us determine whether the M2-V2 registration mark etch at Mask E2 will be concern post E2. Along with the mask interaction concern

978-1-5386-1202-6/19 $31.00 © 2019 IEEE

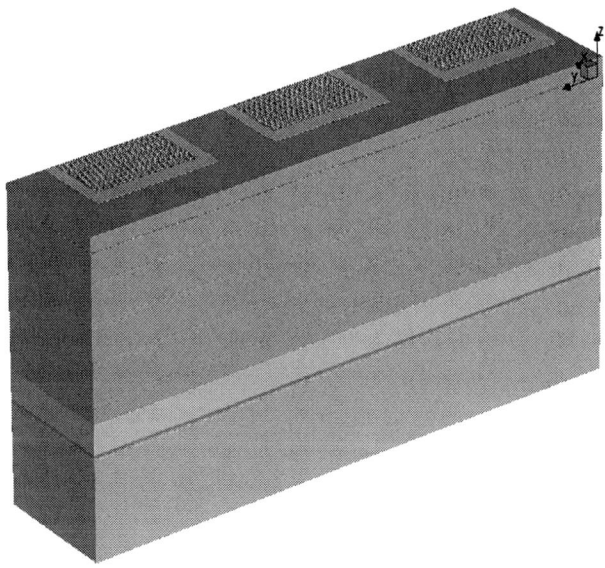

Fig. 4. 3D simulation results for alignment mark highlighted by Mask DRC's

this simulation results can help us determine the aspect ratio of the registration mark and help us determine if there is a topology concern resulting in lifting.

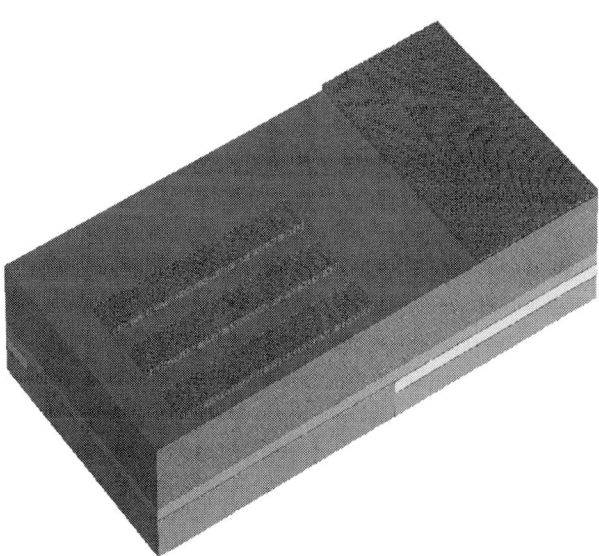

Fig. 5. 3D simulation results for registration mark highlighted by Mask DRC's

From the results obtained in fig. 4 and fig. 5 we can clearly see that the proposed method successfully identifies the areas of concern and predicts the outcome on wafer. So, this pre-silicon verification technique improves the learning time reducing the implementation cost for any new architecture.

IV. CONCLUSION

In this paper a 3D simulation technique to predict the process interaction failure of scribe photo marks has been proposed. This method identifies areas of concern on the scribe and runs structural simulations at those locations to validate the outcome. Areas of concern include layer interactions not duplicated in the die and frame structures differing in magnitude compared to the die. These areas are identified in mask interaction analysis and the 3D simulation provides a step by step process analysis. Simulation results from the proposed method shows that these unintended process interactions can be identified at the pre-silicon verification improving the learning time and reducing the implementation cost for new architectures.

REFERENCES

[1] Z. Or-Bach,"A 1,000x improvement in computer systems by bridging the processor-memory gap," 2017 IEEE SOI-3D-Subthreshold Microelectronics Technology Unified Conference (S3S), Burlingame, CA, 2017, pp. 1-4.

[2] S. Lee, "Scaling trends and challenges of advanced memory technology," Technical Papers of 2014 International Symposium on VLSI Design, Automation and Test, Hsinchu, 2014, pp. 1-1.

[3] G. Moore, J. Liao, S. McDade and B. Verzi, "Accelerating 14nm device learning and yield ramp using parallel test structures as part of a new inline parametric test strategy," Proceedings of the 2015 International Conference on Microelectronic Test Structures, Tempe, AZ, 2015, pp. 44-49.

[4] L. DeBruler et al., "Addressable test structure design enabling parallel testing of reliability devices," 2018 IEEE International Conference on Microelectronic Test Structures (ICMTS), Austin, TX, 2018, pp. 137-141.

[5] F. Khatkhatay et al., "Impact of scribe line (kerf) defectivity on wafer yield," 2018 29th Annual SEMI Advanced Semiconductor Manufacturing Conference (ASMC), Saratoga Springs, NY, 2018, pp. 374-378.

[6] S. Park, "Technology Scaling Challenge and Future Prospects of DRAM and NAND Flash Memory," 2015 IEEE International Memory Workshop (IMW), Monterey, CA, 2015, pp. 1-4.

[7] J. Lin, Burn, "Making lithography work for the 7-nm node and beyond in overlay accuracy, resolution, defect, and cost". Microelectronic Engineering, 2015.

[8] L. Zhang et al., "New alignment mark design structures for higher diffraction order wafer quality enhancement" Vol 10145 SPIE ADVANCED LITHOGRAPHY, Metrology, Inspection, and Process Control for Microlithography XXXI 2017.

Gap in pagination due to withheld paper.

Pages 18-22

Contributed Paper

Resistance-Based Embedded Non-Volatile Memories as Synaptic Devices in Spiking Neural Networks

Foroozan S. Koushan
Department of Electrical and
Computer Engineering,
UC Santa Cruz,
Micron Semiconductor*
fkoushan@micron.com

Sung-Mo Steve Kang
Department of Electrical and
Computer Engineering,
UC Santa Cruz,
skang@ucsc.edu

Abstract—**This paper presents prospect and challenges of using state-of-the-art memory technologies in neuro-inspired computing. As it has become critical to move beyond the Von-Neumann architecture, emulation of biological system has been sought after. In this paper, characteristics of a desired synaptic device required for constructing neural network are examined to highlight what functionalities of synapses can be emulated with new non-volatile memory (NVM) devices, what the basic requirements are to realize artificial inorganic neurons and synapses, and which material and device structures can be used for this purpose.**

Keywords— *Neural network, Von-Neumann architecture, emerging NVM, PCM, RRAM, CBRAM, Memristor, Neuromorphic technology, Synaptic device*

I. INTRODUCTION

In recent years, artificial neural networks (Supervised Learning or Deep Learning) has significantly enhanced accuracy in performing large-scale visual/auditory recognition and classification tasks, some even going beyond human-level accuracy [1]. Convolutional neural network (CNN) [2] and recurrent neural network (RNN) [3] algorithms, in particular, have also improved their capability in the area of video, image, speech, and biomedical applications by increasing the size of data set, implementing better modeling schemes, and combining both. Deep learning algorithms have also helped improve accuracy by aggressively increasing the depth and size of the neural network. But the volume of computations imposes increasingly more significant challenges for hardware implementation of these systems[4].

Today, most of the deep learning training is done by GPUs and some application-specific accelerators [5,6,7]. These accelerators mostly use SRAM as a synaptic memory on-chip, which is extremely inadequate for storage of large amount of data required in deep leaning algorithms. As an alternative solution, emerging NVM devices are considered for on-chip weight storage in parallel computation with low leakage, low power consumption and high density[8].

One of the critical requirements in selecting an NVM device is the device's capability for multi-level resistance or conductance states which would mimic bio-synaptic device characteristics in neural networks. Among emerging NVM devices, Phase Change Memory (PCM), Resistive RAM or Memristors, Spin-Transfer Torque Magnetic RAM, and Floating Gate Memory devices have potential to fulfill this requirement, with minimum compromise in parallel computation of weighted sum.

Many comprehensive research outcomes have been published in the last few years mainly focusing on changes in conductance, or dielectric constant of various metal oxides between two electrodes to emulate basic characteristics of biological synapses at device levels [9, 13, 14] and even in arrays[10, 11, 12]. Intrinsic shortcomings of devices are mitigated by implementing elaborate device architecture and software algorithms to maximize efficiency of the whole system. In addition, there have been some papers attempting to highlight the key characteristics of NVM devices for neural network applications [15], focusing on energy-efficient computation. This paper aims at to discuss the findings on component-level search for best synapses in device and material level, which can be to emulate the realistic biological behavior of synapses in terms of plasticity and accuracy.

II. NEUROMORPHIC DESIGN

A. Review of the hardware design

Current hardware design for neuromorphic networks depends on whether training is done off-line or in-situ in the system. Off-line training is used for non-spiking hardware where information is loaded on edge devices as a one-time programming using software, and classification task is done afterwards by hardware.

In-situ training can be done by two different methodologies: 1) Deep learning which uses supervised techniques to train hardware, based on error calculation of the results vs. real model; and 2) Spiking which uses non-supervised technique for hardware training, using STDP (Spike-Timing Dependent

Plasticity) mimicking the biological process of weighted summing in synapses.

Based on the STDP learning rule, the synapse's conductance or weight decreases if pre-synaptic neuron fires earlier than the post-synaptic neuron, and vice versa. The magnitude of the change in weight depends also on the timing between firing of the two neurons: shorter timing results in larger change.

Emerging NVMs have proved to be a viable replacement for SRAM in GPUs due to their low leakage and potential for parallelism and low power consumption. For back propagation and non-spiking learning, the amount of transferred data increases significantly and power consumption to transfer the data back and forth between the memory and control unit increases exponentially [15]. But, to our best knowledge, for the STDP learning rule eNVMs have not been extensively studied or compared with each other and against bio-synapses.

In the next section, the main characteristics needed for STDP in an ideal synaptic device are summarized, and then compared against some of most promising eNVMs in the market. The main eNVMs considered for this comparison are resistive memories, such as Phase Change Memory(PCM), Resistive RAM (ReRAM), Conductive Bridge RAM (CBRAM), and Spin-Transfer Torque Magnetic RAM (STT-MRAM). Many of these devices have proved to be viable for replacing current flash memory as reported by Sony, Toshiba, Samsung, Micron [17, 18, 19].

B. Synaptic Devices

In this section, desirable characteristics of an ideal synaptic device for STDP application are discussed. The basic characteristics of a synapse can be replicated by using about ten transistors. However, given the fact that the human brain has approximately 10^{11} neurons and 10^{15} synapses, artificial neurons and synapses built on CMOS transistors are unlikely to accommodate the scalability required for neuromorphic computing. Thus, multi-level state capability satisfying the plasticity requirement, accuracy, and linearity to fulfill analog behavior of weight change in biological synapses are crucial characteristic requirements.

Different techniques such as binarization of the neural network parameters by trading off precision of weights to area [20] have been implemented in order to overcome the lack of ideal plasticity requirement in synaptic devices. Also a combination of write voltage and amplitude algorithms have shown to improve nonlinearity/asymmetric characteristic of synapses [29], at the cost of the area and complexity of the digital circuits. But the search to find an ideal synaptic device that performs learning and memory processes at the network level continues.

III. EMERGFING NV MEMORY TECHNOLOGIES

Resistance-based emerging NVMs have shown a great potential to be used as synaptic devices due to their scalability, low power consumption and multi-level state variability.

Among this group are PCM, RRAM/CBRAM, and STT-MRAM.

Scalability and low power consumptions in STT-RAM have made it a strong candidate to replace current flash process compared to other eNVMs. Research at IMEC has shown that by switching to perpendicular Magnetic Tunnel Junction structure (pMTj), STT-RAM can scale down to 10nm and below [31]. Also research has shown that when STT-RAM is used in a non-conventional regimes, it acts as a stochastic memristor, which can implement synaptic functions [24]. But high write energy and long write latency are the most recent challenges in STT-RAM, which industry is trying to overcome by using architectural approaches.

PCM relies on interchangeable crystallization and amorphization of a chalcogenide layer, and achieves multi-level resistance states by modulating the volume of the amorphous state. Gradual device conductance state up to 100 levels have been shown [32]. But the realization of PCM as synaptic device has been difficult so far due to abrupt characteristics of its RESET state compared to SET. This asymmetric characteristics has proved hard to be mitigated by changes in programming algorithms.

Similar problem is observed in RRAM and CBRAM devices which inhibits a non-symmetric characteristics in SET and RESET operations, limiting multi-level state capability per cell and the stability of these states. CBRAM is a version of RRAM where the resistance change is due to the migration of cations from one or both electrodes in the oxide layer, rather than the ionized oxygen or their respective vacancy as in RRAM. Due to the abrupt nature of SET operation in these devices, many tradeoffs have been made in order to use CBRAM and RRAM as synapses for STDP applications. Tradeoffs such as reducing ON and OFF window by limiting the region of ON state and controlling the amount of current flowing through the filament, which results in a gradual increase of conductance in analog form[16]; or adjusting pulse amplitude incremental steps during SET and RESET in RRAM with Pt/HfOx/TiN stack, that compromises speed for accuracy and adds to the latency of write operation[22]. Also recent research shows a tradeoff between the switching speed and the volatile behavior of RRAM by engineering a low mobility metal ions such as Ti within the active electrode[30], and a tradeoff between device footprint and its symmetric switching characteristics by connecting one RRAM to two CMOS transistors[30].

A. Metal-Oxide bilayer Memristor

Material engineering of stack in RRAM and Memristors have opened the path to overcome their basic challenges, like capability of storing multiple memory states per cell and long-term stability of these independent states, mainly for Spiking Neural Networks application. It has been shown that the interface between active layer and one of the electrodes influences the stability and switching characteristics in these devices [25], as it controls the location of oxygen vacancy layer formed during switching. In particular, most stable switching characteristics are observed when the active layer is deposited on top of bottom electrode in a bilayer oxide stack.

Also the number of stable memory states is proved to be modulated by the types of bi-layer oxides in ReRAM devices [26]. Insertion of a thin interfacial layer between the active layer and the contact is proved to be critical in the stability of the filament. Fig. 1 demonstrates how TiO_2 (active layer)+ Al_2O_3(thin barrier) stack results in the highest number of memory states, whereas TiO_2+SiO_2 has the least number of levels[26].

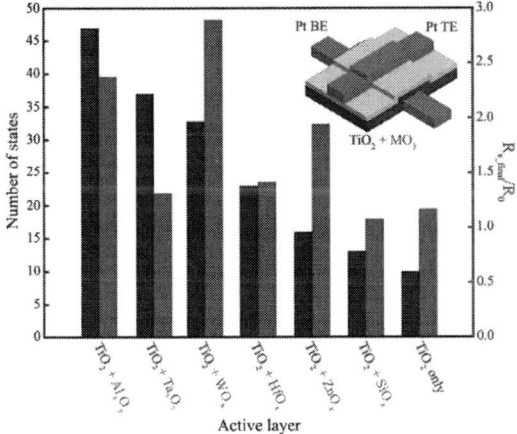

Fig 1. "Bilayer Device statistics. Multibit evaluation of devices based on different barrier layer combinations. Number of attainable resistive states (left axis) and ratio of the final state resistance over the baseline resistance (right axis) for typical bilayer devices". [26]

Incorporation of a thin oxide barrier in Memristor and ReRAM devices could also improve the plasticity of the device. This is an important requirement needed in synaptic devices to function like a chemical synapse to enable learning. In Spiking Neural Network, the final connection between the neurons can be made by high conductive paths generated within the active layer, after training process has modulated the resistance of the barrier layer.

B. Future Work

Since the conductive path formation in a bilayer oxide Memristor devices plays a major role in training process in Spiking Neural Network, in our future work we intend to study this phenomenon, using computational phase-field analysis. In contrast to molecular dynamics which tracks the motion of each charge carrier, the phase-field formulation tracks the dynamical evolution of the envelope of clusters of charge carriers, the aggregate boundary of which forms a conducting channel interface within the non-conducting layer. This investigation is intended to give an understanding on how resistive switching works in a bi-layer oxide, as its behavior depends on dynamical characteristics of atomic metastable states within the thin dielectrics. Conductive path formation and its stability under thermal diffusion will be another aspect in our future investigation.

IV. SUMMARY

The possibility of neuro-inspired computing with eNVMs has increased drastically within the last decade as these devices have proved to have required characteristics to be used as synapses to bring together memory and computational process in the network. Linearity and scalability characteristic of resistance-based eNVMs have made them strong candidates to be used as synaptic devices. Memristors with bilayer oxide stack are demonstrated to have increased number of multi-level states, with long-term stability, making them strong candidates to be used as synaptic devices in STDP. Computational analysis of conductive path formation in these devices will be done in the near future by our team, using Phase Field simulation, for computational analysis of conductive path formation under different thermal conditions.

REFERENCES

[1] Y. LeCun, Y. Bengio, G. Hinton, "Deep learning," Nature , vol. 521, p. 436–444, 2015.

[2] A. Krizhevsky, I. Sutskever, G. E. Hinton, "ImageNet Classification with Deep Convolutional Neural Networks," in Advances in Neural Information Processing Systems, 2012.

[3] A. Graves, A.-r. Mohamed, G. Hinton, "Speech recognition with deep recurrent neural networks," in IEEE International Conference on Acoustics, Speech and Signal Processing (ICASSP), 2013.

[4] Q. Le, M. Ranzato, R. Monga, M. Devin, K. Chen, G. Corrado, J. Dean, A. Ng, "Building high-level features using large scale unsupervised learning," in International Conference in Machine Learning, 2012.

[5] S. B. Furber, F. Galluppi, S. Temple, L. A. Plana, "The SpiNNaker project," Proceedings of the IEEE , vol. 102, no. 5, pp. 652-665, 2014.

[6] Y-H. Chen, T. Krishna, J. Emer, V. Sze, "Eyeriss: An Energy-Efficient Reconfigurable Accelerator for Deep Convolutional Neural Networks," in IEEE International Solid-State Circuits Conference (ISSCC), 2016.

[7] J. Sim, J -S. Park, M. Kim, D. Bae, Y. Choi, L -S. Kim, "A 1.42TOPS/W Deep Convolutional Neural Network Recognition Processor for Intelligent IoE Systems," in IEEE International Solid-State Circuits Conference (ISSCC), 2016.

[8] S. Yu (Ed), Neuro -inspired Computing Using Resistive Synaptic Devices, Springer, 2017.

[9] D. Kuzum, S. Yu, and H. -S. P. Wong, "Synaptic electronics: materials, devices and applications," Nanotechnology, vol. 24, p. 382001, 2013.

[10] G. Indiveri, et al., "Neuromorphic silicon neuron circuits," Frontiers in Neuroscience, vol. 5, no. 73, pp. 1 -23, 2011.

[11] J. Hasler, and B. Marr, "Finding a road map to achieve large neuromorphic hardware systems," Frontiers in Neuroscience, vol. 7, no. 118, pp. 1 -29, 2013.

[12] S. Furber, "Large -scale neuromorphic computing systems," Journal of Neural Engineering, vol. 13, p. 051001, 2016.

[13] D. S. Jeong, I. Kim, M. Zieglerb, and H. Kohlstedtb, "Towards artificial neurons and synapses: a materials point of view," RSC Advances, vol. 3, p. 3169, 2013.

[14] D. S. Jeong, K. M. Kim, S. Kim, B. J. Choi, and C. S. Hwang, "Memristors for energyefficient new computing paradigms," Advanced Electronic Materials, vol. 2, p. 1600090, 2016.

[15] Shimeng Yu, "Neuro-inspired computing with emerging non-volatile memorys," Proceedings of the IEEE, vol. 106, no. 2, pp. 260-285, 2018

[16] Y. Shi, L. Nguyen, S. Oh, X. Liu, F. Koushan, J. Jameson, D. Kuzum, "Neuro-inspired Unsupervised Learning and Pruning with Subquantum CBRAM Arrays," Nature Communications, 9(1), 5312, 2018

[17] Y. Choi, I. Song, M-H. Park, H. Chung, S. Chang, B. Cho, J. Kim, Y. Oh, D. Kwon, J. Sunwoo, J. Shin, Y. Rho, C. Lee, M. Kang, J. Lee, Y. Kwon, S. Kim, J. Kim, Y-J. Lee, Q. Wang, S. Cha, S. Ahn, H. Horii, J. Lee, K. Kim, H. Joo, K. Lee, Y -T. Lee, et al., "A 20nm 1.8V 8Gb PRAM with 40MB/s program bandwidth," in IEEE International Solid-State Circuits Conference (ISSCC), 2012

[18] T. -Y. Liu, T. H. Yan, R. Scheuerlein, Y. Chen, J. K. Lee, G. Balakrishnan, G. Yee, H. Zhang, A. Yap, J. Ouyang, T. Sasaki, S. Addepalli, A. Al -Shamma, C. -Y. Chen, M. Gupta, G. Hilton, S. Joshi, A. Kathuria, V. Lai, D. Masiwal, M. Matsumoto, et al. , "A 130.7mm2 2-layer 32Gb ReRAM memory device in 24nm technology," in IEEE International SolidState Circuits Conference, 2013.

[19] A. Kawahara, R. Azuma, Y. Ikeda, K. Kawai, Y. Katoh, K. Tanabe, T. Nakamura, Y. Sumimoto, N. Yamada, N. Nakai, S. Sakamoto, Y. Hayakawa, K. Tsuji, S. Yoneda, A. Himeno, K. Origasa, K. Shimakawa, T. Takagi, T. Mikawa, and K. Aono, "An 8Mb multi - layered cross -point ReRAM macro with 443MB/s write throughput," in IEEE International Solid-State Circuits Conference, 2012.

[20] J. Woo, K. Moon, J. Song, S. Lee, M. Kwak, J. Park, and H. Hwang, "Improved synaptic behavior under identical pulses using AlOx/HfO2 bilayer RRAM array for neuromorphic systems," IEEE Electron Device Letters, vol. 37, no. 8, pp. 994 -997, 2016.

[21] S. Yu, B. Gao, Z. Fang, H. Y. Yu, J. F. Kang, and H. -S. P. Wong, "Stochastic learning in oxide binary synaptic device for neuromorphic computing," Frontiers in Neuroscience, vol. 7, p. 186, 2013.

[22] L. Gao, and S. Yu, "Programming protocol optimization for analog weight tuning in resistive memories," in IEEE Device Research Conference (DRC), 2015.

[23] J. Woo, K. Moon, J. Song, S. Lee, M. Kwak, J. Park, and H. Hwang, "Improved synaptic behavior under identical pulses using AlOx/HfO2 bilayer RRAM array for neuromorphic systems," IEEE Electron Device Letters, vol. 37, no. 8, pp. 994 -997, 2016.

[24] A. F.Vincent, J. Larroque, N. Locatelli, N. B. Romdhane, O. Bichler, Ch. Gamrat, W.S. Zhao, J.O. Klein, S.Galdin, D. Querlioz, "Spin-Transfer Torque Magnetic Memory as a Stochastic Memristive Synapse for Neuromorphic Systems," IEEE Transaction on Biomedical Circuits and Systems, vol. 9, no.2, pp. 166-174, 2015.

[25] Alekseeva, L., Nabatame, T., Chikyow, T. & Petrov, A. "Resistive switching characteristics in memristors with Al2O3/TiO2 and TiO2/Al2O3 bilayers." Jpn. J. Appl. Phys. 55, 08PB02, 2016.

[26] S. Stathopoulos, A. Khiat, M. Trapatseli, S. Cortese, A. Serb, I. Valov, Th. Prodromakis, "Multibit Memory Operation of Metal-Oxide Bi-Layer Memristors," J. Nature. Repport 7, no. 17532, 2017.

[27] Alekseeva, L., Nabatame, T., Chikyow, T. & Petrov, A. "Resistive switching characteristics in memristors with Al2O3/TiO2 and TiO2/Al2O3 bilayers." Jpn. J. Appl. Phys. 55, 08PB02, 2016.

[28] S. Stathopoulos, A. Khiat, M. Trapatseli, S. Cortese, A. Serb, I. Valov, Th. Prodromakis, "Multibit Memory Operation of Metal-Oxide Bi-Layer Memristors," ScientificReports 7, no. 17532, 2017.

[29] P. -Y. Chen, B. Lin, I. -T. Wang, T. -H. Hou, J. Ye, S. Vrudhula, J. -S. Seo, Y. Cao, and S. Yu, "Mitigating effects of non -ideal synaptic device characteristics for on -chip learning," in IEEE/ACM International Conference on Computer -Aided Design (ICCAD), 2015.

[30] D. Ielmini, "Brain_inspired Computing with Switching Memory (RRAM): Devices, Synapses and Neural Networks," Microelectronics Engineering, vol. 190, pp.44-53, 2018

[31] E.Liu, J.Swerts, Y.Ch.Wu, A.Vaysset, "Top-Pinned STT-MRAM Devices with High Thermal Stability Hybrid Free Layers for High-Density Memory Applications," IEEE Transactions on Magnetics, PP. 1-5. 10.1109/TMAG.2018.2831904

[32] D. Kuzum, R. G. D. Jeyasingh, B. Lee, and H.-S. P. Wong, "Nanoelectronic programmable synapses based on phase change materials for brain-inspired computing," Nano Letters, vol. 12, no. 5, p. 2179–2186, 2012

Contributed Paper

Three Dimensional Time Domain Simulation of the Quantum Magnetic Susceptibility

Jennifer Houle, Dennis Sullivan, *Fellow, IEEE*, Ethan Crowell, Sean Mossman, and Mark G. Kuzyk

Abstract— **A way of using the Finite Difference Time Domain method is described to simulate the magnetic susceptibility of a quantum toroid. This simulation is based on the direct implementation of the time-dependent Schrödinger equation in three dimensions. First, the ground state eigenenergy and eigenstate are found. Next, the expectation value of the quantum magnetic dipole operator is calculated as a function of the applied magnetic field strength with a static magnetic field, and the results are compared with classical results. Then the magnetic dipole moment is calculated with a time-oscillating magnetic field applied. These expectation values are used to calculate the linear and nonlinear magnetic susceptibility of a torus, both without a grating and with a grating to increase irregularities in the shape, by repeating the calculations at various frequencies. The results are consistent with the expected results. This method can be used to calculate the quantum magnetic susceptibility of any structure in order to search for structures with better nonlinear properties.**

Index Terms— **Computer simulation, Finite difference methods, Magnetic susceptibility, Nonlinear optics, Quantum mechanics.**

I. INTRODUCTION

MATERIALS with nonlinear properties have applications in optical switching [1], lasers, photovoltaic cells [2], imaging [2], cancer therapy [3, 4], advanced computing [5, 6], and communication technologies [7, 8]. Nonlinear properties allow the interaction between multiple signals. To improve applications, it is important to establish good methods and models to explore the optimization of nonlinear effects by increasing the nonlinear response.

The Finite-Difference Time-Domain (FDTD) method offers a method of directly implementing the Schrödinger equation in a three-dimensional structure and has been applied to quantum simulation [9-13]. The accuracy of this method in determining the eigenstates of quantum wires was previously described [14]. The FDTD method was used in the determination of the hyperpolarizability of quantum wires in close proximity to an electric dipole [15].

This paper will describe a torus structure and show the eigenstate generated using the FDTD implementation of the time-dependent Schrödinger equation [17-19]. The FDTD implementation of the magnetic dipole moment operator will be described [16]. Then a simulation will be presented for a time-harmonic magnetic field, and this will be used to calculate the magnetic susceptibility of a torus structure. This will allow for the optimization of structures to enhance the nonlinear properties in the presence of a time-varying magnetic field. Irregularities in structures, likely provided by etching processes, create a nonlinear response, so a grated torus will also be evaluated.

II. THE FINITE-DIFFERENCE TIME-DOMAIN METHOD AND THE DETERMINATION OF EIGENENERGIES AND EIGENSTATES

The time-dependent Schrödinger equation [19] is given by:

$$\frac{\partial \Psi(x,y,z,t)}{\partial t} = i \frac{\hbar}{2m_e}\left[\frac{\partial^2 \Psi(x,y,z,t)}{\partial x^2} + \frac{\partial^2 \Psi(x,y,z,t)}{\partial y^2} + \frac{\partial^2 \Psi(x,y,z,t)}{\partial z^2}\right] \quad (1)$$
$$-\frac{i}{\hbar}V(x,y,z)\Psi(x,y,z,t).$$

From [19], Ψ contains both a real and an imaginary component, allowing Ψ to be separated and (1) to be written as a pair of coupled equations, which may be run sequentially. These equations are evaluated at each time step for each cell in a three-dimensional array representing the problem space. The alternating of the real and imaginary components allows the simulation of the behavior of Ψ over time. Further details can be found in [9-11, 16, 20]. The Ψ calculations over time can be used to find the eigenstates and eigenenergies [16-19].

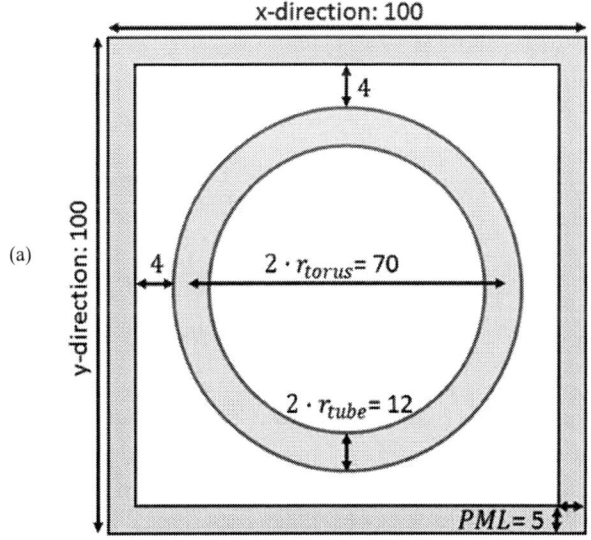

Submitted January 23, 2019. E. C, S. M, and M. G. K. acknowledge the generous support of the National Science Foundation, Grant ECCS-1128076.

J. E. Houle was with the University of Idaho, Moscow, ID, 83843. She is now with is with Moscow-Berlin Simulations in Moscow, ID, 83843 (e-mail:. phil6661@vandals.uidaho.edu).

D. M. Sullivan is with the University of Idaho in Moscow, ID, 83843 (e-mail: dsulliva@uidaho.edu).

E. C, S. M., and M. G. K. are with Washington State University in Pullman, WA, 99164.

978-1-5386-1202-6/19 $31.00 © 2019 IEEE

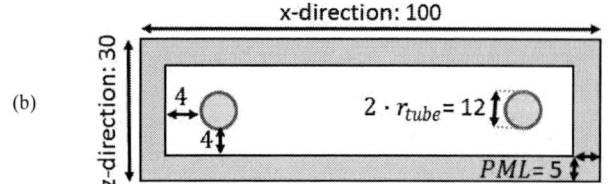

(b)

Fig. 1. The problem space for the torus in the (a) x-y direction, and (b) x-z direction. The total problem space is 100 x 100 x 30 cells, with each cell representing one angstrom cubed. The torus radius is 35 angstroms, and the tube radius is 6 angstroms. The PML boundary is five cells in each direction.

This work will evaluate a torus with a diameter of 70 angstroms and a tube diameter of 12 angstroms. A perfectly matched layer (PML) [21, 22] of five cells surrounds the torus to absorb outgoing waves as described in [13]. Fig. 1 shows the simulation space for the torus. The ground eigenstate for the torus is shown in Fig. 2.

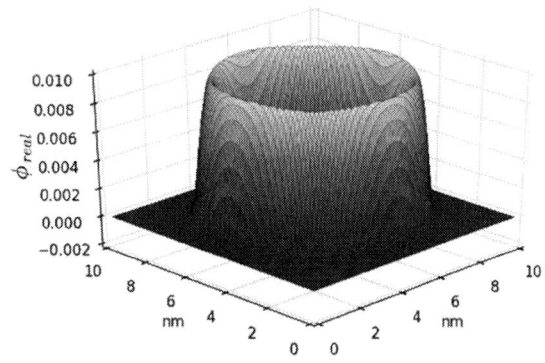

Fig. 2. The ground eigenstate of the torus taken through the center of the torus.

III. CALCULATION OF THE MAGNETIC DIPOLE MOMENT

The Hamiltonian when a magnetic field is applied is given by

$$H = \frac{1}{2m_e}\left(\frac{\hbar}{i}\nabla - q\cdot\mathbf{A}\right)^2 + V(x,y,z), \tag{2}$$

where $\mathbf{A}$ is the vector potential [19]. A static magnetic field B_0 is applied in the z-direction, perpendicular to the torus. Then $\mathbf{A}$ can be simplified to

$$\mathbf{A} = \frac{1}{2}\left[-\left(yB_0\right)\hat{\mathbf{x}} + \left(xB_0\right)\hat{\mathbf{y}}\right]. \tag{3}$$

Ignoring the electric dipole interaction, this results in the Hamiltonian:

$$H = -\frac{\hbar^2}{2m_e}\left(\frac{\partial^2}{\partial x^2}+\frac{\partial^2}{\partial y^2}+\frac{\partial^2}{\partial z^2}\right) + i\frac{\hbar q B_0}{2m_e}\left(-y\frac{\partial}{\partial x}+x\frac{\partial}{\partial y}\right)$$
$$+\frac{q^2 B_0^2}{8m_e}\left(x^2+y^2\right)+V(x,y,z). \tag{4}$$

From [24], the magnetic dipole moment operator is

$$\mathbf{m} = -\frac{1}{2}\frac{q}{m_e}\frac{\hbar}{i}\left(\mathbf{r}\times\nabla\right)-\frac{q^2}{4m_e}B_z(t)\cdot\left(x^2+y^2\right). \tag{5}$$

This is then implemented in FDTD, with the x- and y- positions

being taken from the center of the torus. The details of the FDTD implementation can be found in [16].

The accuracy of the expectation value of the magnetic dipole moment can be verified by comparing it with the classical magnetic dipole moment given by

$$\mathbf{m} = I\cdot\pi\cdot r_{torus}^2\,\hat{\mathbf{z}}, \tag{6}$$

where I is the current and r_{torus} is the radius of the torus [23]. A wave-packet within a radial Gaussian envelope is initialized in the torus and assigned the charge of one electron. This is initialized as shown in Fig. 3 at $t = 0$ fs. The wave-packet will travel around the torus as the FDTD simulation runs, and a complete circle will represent a current $I = q/T$, where T is the time it takes to travel a complete circle.

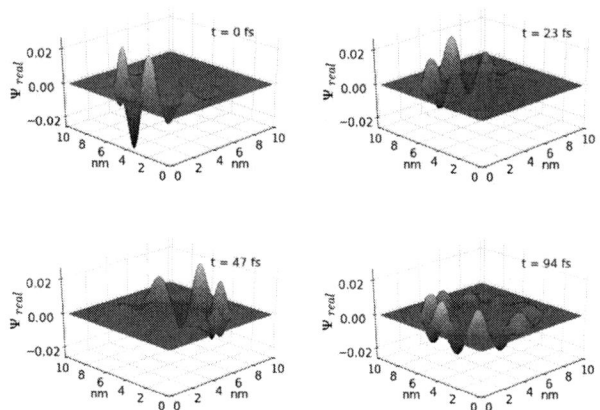

Fig. 3. Progression of a wave packet over time within a torus with a 0T magnetic field applied

Since the wave-packet is initiated to move around the torus, it has no initial momentum in the z-direction and only the x and y expectation values of position are monitored. The amount of time taken to complete a cycle with 0T applied is 84.0 fs. This allows the current to be found, and the current is used to calculate the magnetic dipole moment using (6).

Table 1 shows the results of the classical and FDTD quantum approach for several magnetic field strengths. The classical and quantum approaches give values that are the same within about three percent, verifying the magnetic dipole moment operator and the magnetic field implementation using the FDTD method.

Table 1. Classical and quantum magnetic dipole moment with a static magnetic field

Magnetic Field [T]	Classical $\mathbf{m}\left[mA\cdot\left(\text{Å}\right)^2\right]$	Operator $\mathbf{m}\left[mA\cdot\left(\text{Å}\right)^2\right]$	Difference
-50	7.8	7.6	2.6%
-25	7.5	7.4	1.3%
0	7.3	7.2	1.4%
25	7.1	7.0	1.4%
50	6.9	6.8	1.5%

978-1-5386-1202-6/19 $31.00 © 2019 IEEE

IV. SIMULATION OF A TIME-HARMONIC MAGNETIC FIELD

The applied magnetic field is a sinusoidal function of the form

$$\mathbf{B}(t) = B_{max} \cdot \sin(2\pi f_0 t)\hat{\mathbf{z}}. \qquad (7)$$

This will be multiplied by a Hanning window to reduce the effects of the abrupt addition of a magnetic field. The effects of the magnetic field may be monitored by observing the magnetic dipole moment using the operator described in (5).

Fig. 4(a) shows an applied 30 T, 20 THz magnetic field over time. Fig. 4(b) shows the calculated magnetic dipole moment over time. Fig. 4(c) and Fig. 4(d) show the Fourier transforms of the magnetic field and the magnetic dipole moment, respectively. The peak for the magnetic field is 30 T at 20 THz in Fig. 4(c), as expected given the input magnetic field. The magnetic dipole moment shown in Fig. 4(d) also shows a peak at 20 THz and no significant response at other frequencies.

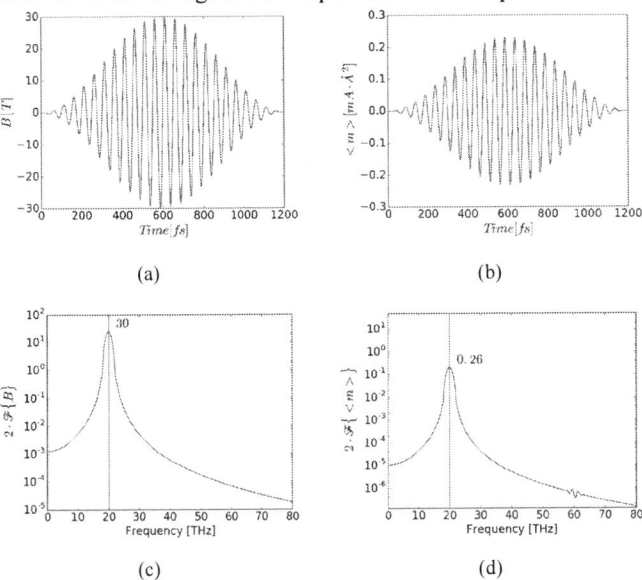

(a)

(b)

(c)

(d)

Fig. 4. The results of a simulation of a particle in a torus under the influence of a sinusoidal magnetic field with a maximum magnitude of 30 T and a frequency of 20 THz. (a) The applied magnetic field over time, (b) the calculated magnetic dipole moment over time, (c) the Fourier transform of the magnetic field showing the applied magnetic field strength of 30 T at 20 THz, and (d) the Fourier transform of the dipole moment showing the primary response is at the applied frequency of 20 THz.

The peak values of the magnetic dipole moment generated with the Fourier transform are determined at f_0. These are shown to have the expected linear relationship with the applied magnetic field strength.

V. MAGNETIC SUSCEPTIBILITY

The magnetic dipole moment generated from the FDTD method using the Fourier transform, as in Fig. 4(d), can be used to find the magnetic susceptibility.

From [1], the bulk polarizability is given by

$$P(t) = \chi^{(1)}E(t) + \chi^{(2)}E^2(t) + \chi^{(3)}E^3(t) + \ldots \qquad (8)$$

This response is in the z-direction only, which is the direction of the applied magnetic field. The bulk polarizability is related to the dipole moment of a molecule by

$$P(t) = N \cdot p(t), \qquad (9)$$

where N is the number density and certain conditions are met as described in [1]. The dipole moment equation may then be written as

$$p(t) = \alpha E(t) + \beta E^2(t) + \gamma E^3(t) + \ldots \qquad (10)$$

An analogous equation for the magnetic dipole moment m using the magnetic field $B(t)$ results in

$$m(t) = \alpha'B(t) + \beta'B^2(t) + \gamma'B^3(t) + \ldots \qquad (11)$$

where the primed quantities are the linear and nonlinear electric polarizabilities and hyperpolarizabilities. The following calculations will focus on α (the magnetic polarizability), β (the magnetic hyperpolarizability), and γ (the magnetic second hyperpolarizability) [2]. The response is needed in the frequency domain so the magnetic dipole moment is written as

$$m(\omega) = \alpha'B(\omega) + \beta'B^2(\omega) + \gamma'B^3(\omega). \qquad (12)$$

A Fourier transform is performed to obtain the magnetic dipole moment in the frequency domain.

The polarizability will occur at the fundamental frequency, ω_0. The time varying magnetic field was previously described in (7). This means the hyperpolarizability will have a magnetic field component

$$B^2(t) = \frac{1}{2}\left(B_{max}^2 - B_{max}^2\cos(2\omega_0 t)\right), \qquad (13)$$

which has a frequency component at 0 and at $2\omega_0$ [25].

Similarly, the second hyperpolarizability will have a magnetic component

$$B^3(t) = \frac{1}{4}\left(3B_{max}^3\sin(\omega_0 t) - B_{max}^3\sin(3\omega_0 t)\right). \qquad (14)$$

This means the second hyperpolarizability will have frequency components at ω_0 and $3\omega_0$.

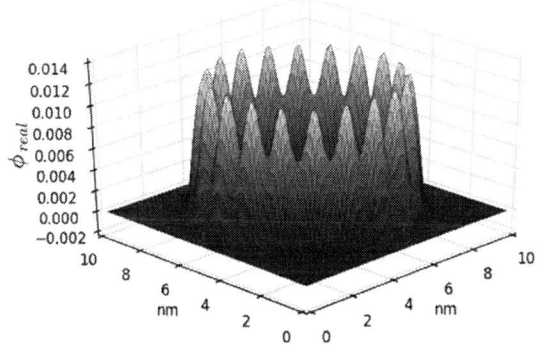

Fig. 5. The ground eigenstate for the grated potential with the cross section shown at z = ZC. This ground eigenstate will be used for the initial value when applying the time harmonic magnetic field.

The Fourier transform of the simulation results for the plain torus in Fig. 4(d) showed only a small component at $3f_0$. In an attempt to find a higher nonlinear response, irregularities were added to the structure. A grated torus, one in which a periodic potential is added, is used in this work. This structure has 20 peaks. Any structure may be examined using the FDTD method simply by specifying the potential $V(x,y,z)$. The ground eigenstate for the grated torus is shown in Fig. 5.

The results of a magnetic field described by (7) where $B_{max} = 30$ T and $f_0 = 20$ THz on the grated torus are shown in Fig. 6. The majority of the signal for the magnetic dipole moment is at f_0, but there is also a significant component at $3f_0$, as shown in Fig. 6(d). This corresponds with the second hyperpolarizability. This component is much more prominent than that seen in the plain torus in Fig. 4(d).

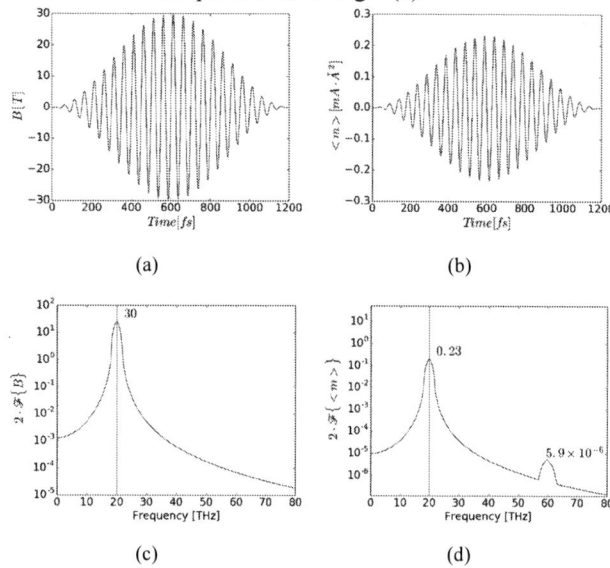

(a)　　　　　　　　(b)

(c)　　　　　　　　(d)

Fig. 6. The results of a simulation of a particle in a grated torus under the influence of a sinusoidal magnetic field with a maximum magnitude of 30 T and a frequency of 20 THz. (a) The applied magnetic field over time, (b) the calculated magnetic dipole moment over time, (c) the Fourier transform of the magnetic field showing the applied magnetic field strength of 30 T at 20 THz, and (d) the Fourier transform of the dipole moment showing the primary response is at the applied frequency of 20 THz and a secondary response at 60 THz.

Fig. 7 shows the component of the magnetic dipole moment determined at 20 THz, f_0, across several magnetic field strengths. Because the hyperpolarizability has a cubic component at f_0, as shown in (14), a best fit line is created using a linear and a cubic response to match the calculated values. The linear coefficient, α, is dominant and gives the polarizability. The cubic coefficient, γ, is a component of the second hyperpolarizability. These coefficients are used to create the solid fit line on the graph. The cubic term is calculated for comparison with the cubic fit that will be found at $3f_0$ but, being several orders of magnitude smaller than the linear term, is not visibly apparent in Fig. 7.

Similarly, the magnetic dipole moment from the Fourier transform can be determined at $2f_0$ to examine the hyperpolarizability. In the case of both the plain and grated torus, this component is not present in any significant amount.

Fig. 8 shows the component of the magnetic dipole moment at $3f_0$. The coefficient of the cubic response is a component of γ, the second hyperpolarizability. The cubic fit line is also

shown in Fig. 8. The ratio of γ determined at $3f_0$ to γ determined at f_0 is approximately -0.35, which is close to the ideal value of -1/3. This ratio can be seen in (14).

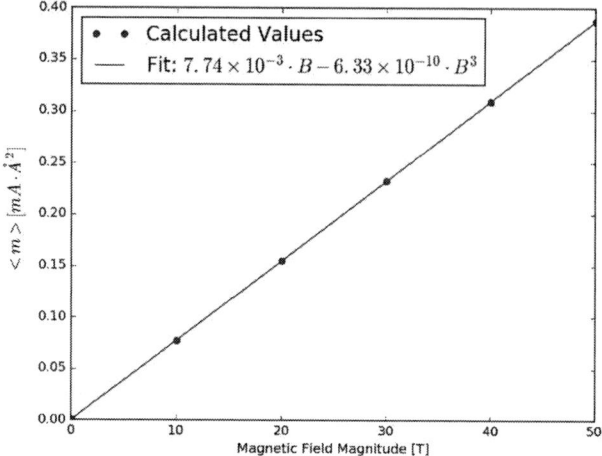

Fig. 7. The absolute value of the magnetic dipole moment determined at f_0 versus magnetic field strength for a particle in a grated torus. A best fit line is used to find the linear coefficient ($\alpha = 7.74 \times 10^{-3}$ mA $\cdot$ Å$^2 \cdot$ T^{-1}) and the cubic coefficient ($\gamma = -6.33 \times 10^{-10}$ mA $\cdot$ Å$^2 \cdot$ T^{-3}).

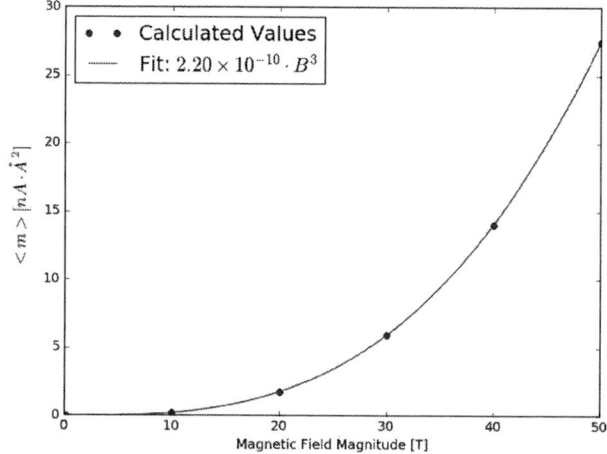

Fig. 8. The absolute value of the magnetic dipole moment determined at $3f_0$ versus magnetic field strength. A best fit line is used to find the cubic coefficient ($\gamma = 2.20 \times 10^{-10}$ mA $\cdot$ Å$^2 \cdot$ T^{-3}).

VI. CONCLUSION

The FDTD method can simulate any structure, subject only to the resolution of the cell size and the dimensions of the problem space. This allows different nanostructures to be simulated, tested, and modified to search for the optimal nonlinear optical response. Future work will focus on creating these optimized structures.

REFERENCES

[1] R.W. Boyd, *Nonlinear Optics, Second Edition*. Academic Press, Cambridge, MA, USA: 2003.

[2] J. Pérez-Moreno, S. Shafei, M. G. Kuzyk, "Sum Rules and Scaling in Nonlinear Optics," *Physics Reports*, vol. 524, no. 4, pp. 397-398. April 2013.

[3] M. Firczuk, M. Winiarska, A. Szokalska, M. Jodlowska, M. Swiech, K. Bojarczuk, P. Salwa, D. Nowis, "Approaches to improve photodynamic therapy of cancer," *Front Biosci, Landmark 16*, pp. 208-224. 2011.

[4] I. Roy, T. Y. Ohulchanskyy, H. E. Puda. J. Bergey, A. R. Oseroff, J. Morgan, T. J. Dougherty, and P. N. Prasad, "Ceramic-Based Nanoparticles Entrapping Water-Insoluble Photosensitizing Anticancer Drugs: A Novel Drug–Carrier System for Photodynamic Therapy," *J. Am. Chem. Soc*, vol. 125, no. 26, pp. 7860–7865. 2003.

[5] D. J. Brod, and J. Combe. "Passive CPHASE gate via cross-Kerr nonlinearities," *Phys. Rev. Lett.* vol. 117, 080502. Aug. 2016.

[6] D. J. Brod, J. Combes, J. Gea-Banacloche. "Two photons co- and counterpropagating through N cross-Kerr sites," *Phys. Rev. A*, vol. 94: 023833. Aug. 2016.

[7] D. Cotter, R. J. Manning, K. J. Blow, A. D. Ellis, A. E. Kelly, et al. "Nonlinear optics for high-speed digital information processing," *Science,* vol. 286: pp. 1523-1528, Nov. 1999.

[8] J. E. Heebner and R. W. Boyd, "Enhanced all-optical switching by use of a nonlinear fiber ring resonator," *Opt Lett*, vol. 24: 847-849. 1999.

[9] D. M. Sullivan and D. S. Citrin, "Time-domain simulation of two electrons in a quantum dot," *J. Applied Physics*, vol. 89: 3841-3846. 2001.

[10] D. M. Sullivan and D. S. Citrin, "Time domain simulation of quantum spin," *J. Applied Physics*, vol. 94: pp. 6518. 2003.

[11] A. Soriano, E. A. Navarro, J. A. Porti, and V. Such, "Analysis of the finite difference time domain technique to solve the Schrödinger equation for quantum devices," *J. Applied Physics*, vol. 95: 8011-8018. 2004.

[12] G. B. Ren and J. M. Rorison, "Eigenvalue problem of the Schrödinger equation via the finite-difference time-domain method," *Phys Rev E Stat Nonlin Soft Matter Phys*, vol. 69: 036705. Mar. 2004.

[13] D. M. Sullivan and P. M. Wilson, "Time-domain determination of transmission in quantum nanostructures," *J. Applied Physics*, vol. 112: 064325. 2012.

[14] D. M. Sullivan, S. Mossman, and M. Kuzyk, "Time-domain simulation of three-dimensional quantum wires," PLOS One: 0153802. April 2016.

[15] D. M. Sullivan, S. Mossman, and M. Kuzyk, "Hybrid quantum systems for enhanced nonlinear optical susceptibilities," *J. Opt Soc Am B,* vol. 33: E143-E149. 2016.

[16] J. Houle, D. M. Sullivan, E. Crowell, S. Mossman, M. G. Kuzyk, "Three dimensional time domain simulation of the quantum magnetic dipole," *International Journal of Magnetics and Electromagnetism.* 4: 011. 2018.

[17] D. M. Sullivan and D. S. Citrin, "Determination of the eigenfunctions of arbitrary nanostructures using time domain simulation," *J Applied Physics*, vol. 91: pp. 3219-3226. 2002.

[18] D. M. Sullivan and D. S. Citrin, "Determining quantum eigenfunctions in three-dimensional nanoscale structures," *J. Applied Physics*, vol. 97: 104305. 2005.

[19] D. M. Sullivan, *Quantum mechanics for electrical engineers*. IEEE Press. 2012.

[20] W. Dai, G. Li, R. Nassar, and S. Su, "On the stability of the FDTD method for solving a time-dependent Schrödinger Equation," *Numerical Methods for Partial Differential Equations*, vol. 21: 1140-1154. April 2005.

[21] J. P. Berenger, "A perfectly matched layer for the absorption of electromagnetic waves," *J. Comput Phys*, vol. 114: pp. 185-200. Oct. 1994.

[22] C. Zheng, "A perfectly matched layer approach to the nonlinear Schrödinger wave equations," *J. Compt Phys*, vol. 227: pp. 537-556. Nov. 2007.

[23] D. K. Cheng, *Field and Wave Electromagnetics*, Addison-Wesley Publishing, 1989.

[24] A. L. Fetter and J. D. Walecka, *Theoretical Mechanics of particles and continua*, Dover Publishing, New York. 2003.

[25] M. G. Kuzyk, *Nonlinear Optics: A Student's Perspective with python problems and examples*. Createspace. 2018.

978-1-5386-1202-6/19 $31.00 © 2019 IEEE

Contributed Paper

An optically reconfigurable gate array workable under a strong gamma radiation environment

Shinya Fujisaki, Takumi Fujimori, and Minoru Watanabe

Electrical and Electronic Engineering
Shizuoka University
3-5-1 Johoku, Hamamatsu, Shizuoka 432-8561, Japan
Email: watanabe.minoru@shizuoka.ac.jp

Abstract— **Optically reconfigurable gate arrays (ORGAs) have been under development for use as radiation-hardened devices. The ORGAs are a type of field programmable gate array (FPGA). However, by introducing an optical holographic memory technology onto currently available semiconductor technology, the radiation tolerances of ORGAs could be increased drastically. The total-ionizing-dose tolerance of the ORGA VLSI has reached over 400 Mrad, which is 400 times higher radiation tolerance than current radiation-hardened VLSIs. Moreover, we have developed a radiation-hardened power supply unit based on lithium-ion battery cells. For this experiment, we have demonstrated that an ORGA powered by the radiation-hardened power supply unit can function correctly for 24 hr at a dose rate of 15.3-22 rad/s (Si) using a Co60 gamma radiation source.**

I. INTRODUCTION

Decommissioning operations are scheduled for the Fukushima Daiichi nuclear power plant which was destroyed by an earthquake and its related tsunami disaster on March 2011 [1][2]. Since some tasks of decommissioning operations in the Fukushima Daiichi nuclear power plant must be done under 1000 Sv/h intense radiation conditions in the worst case, more robust radiation-hardened devices than currently available radiation-hardened devices with up to a 1 Mrad total-ionizing-dose tolerance are required [3][4].

Therefore, we have been developing a new radiation-hardened optically reconfigurable gate array (ORGA) used for robots performing tasks for decommissioning operations [5]-[9]. The ORGA comprises a holographic memory, a laser array, and an ORGA VLSI, which includes a fine-grained programmable gate array and field programmable gate arrays (FPGAs). To date, the radiation-hardened ORGA system has been demonstrated to have a 400 Mrad total-ionizing-dose tolerance, which is 400 times higher than currently available radiation-hardened VLSIs.

Moreover, we have been developing a radiation-hardened power supply unit based on lithium-ion battery cells. The radiation tolerance has been confirmed by using a Co60 gamma radiation source.

This paper presents the demonstration result that an ORGA powered by the radiation-hardened power supply unit can function correctly for 24 hr at a dose rate of 15.3-22 rad/s (Si).

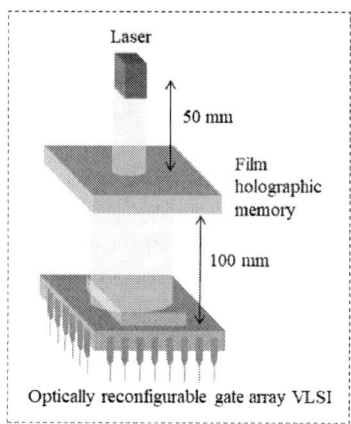

Fig. 1. Block diagram of an ORGA consisting of a laser, a film holographic memory, and an ORGA-VLSI.

TABLE I

CHIP CHARACTERISTICS

Technology		0.18μm double-poly 5-metal standard CMOS process
Die size		5.0×5.0 mm^2
Supply voltage	Core	1.8V
	I/O	3.3V
Photodiode	Junction area	4.40×4.45μm^2
	Switching energy	2.12×10^{-14}J
	Horizontal interval	30.08μm
	Vertical interval	30.24μm
	No. of photodiodes	17,664
Gate array	No. of logic blocks	128
	No. of switching matrices	144
	No. of Wires in a wiring channel	8
	No. of I/O blocks	16 (64bit)
	Gate count	8,704

II. RADIATION-HARDENED OPTICALLY RECONFIGURABLE GATE ARRAY (ORGA)

A radiation-hardened ORGA consists of a holographic memory, a laser array, and an ORGA-VLSI. A block diagram is presented in Fig. 1. The ORGA-VLSI includes a fine-grained programmable gate array that is similar to current field programmable gate arrays (FPGAs).

978-1-5386-1202-6/19 $31.00 © 2019 IEEE

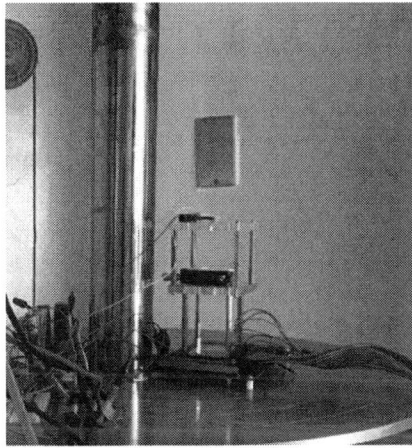

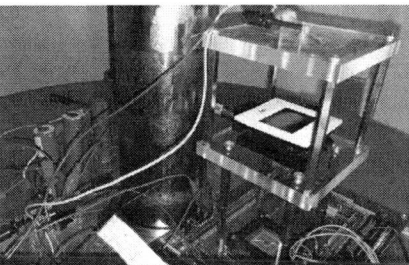

Fig. 2. Photograph of a radiation experiment using the ORGA comprising a laser, a holographic memory, and an ORGA VLSI at a dose rate of 85 rad/s (Si) using a Co60 gamma radiation source.

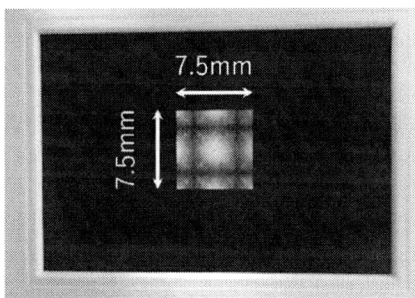

Fig. 3. Film holographic memory recording an AND circuit.

The specifications are presented in Table I. The ORGA-VLSI includes 128 logic blocks and 144 switching matrices. Each logic block includes two 4-input look-up tables and two D-type flip-flop. The ORGA-VLSI has 17,664 photodiodes used for optical configuration. Current FPGAs use a serial configuration architecture. However, the serial configuration architecture is exceedingly vulnerable to radiation. Therefore, the ORGA-VLSI has introduced a parallel configuration architecture that includes many photodiodes, as shown in Fig. 6, so that the ORGA-VLSI can achieve a perfectly parallel configuration. Each programming point of the programmable gate array on the ORGA-VLSI is connected to a photodiode. Even if a part of the programmable gate array and a part of the configuration function of the

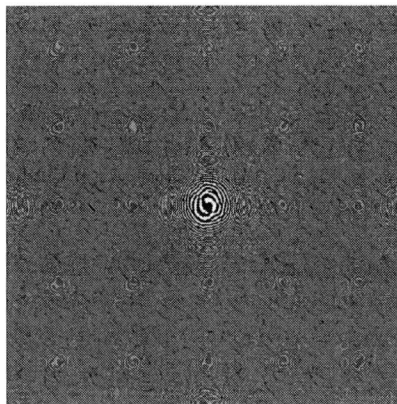

Fig. 4. Binary holographic memory pattern consisting of 1,200 x 1,200 pixels recorded on the film holographic memory shown in Fig.3.

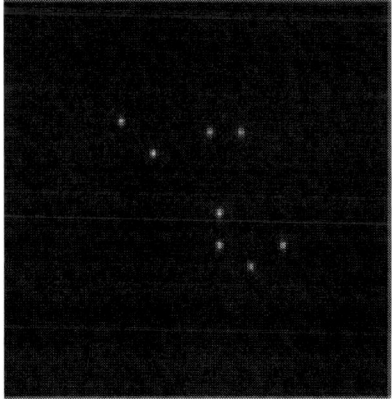

Fig. 5. CCD-captured configuration context pattern of the AND circuit.

programmable gate array are damaged by radiation, the non-damaged rest region of the programmable gate array can be programmed correctly using optical configuration.

Such optical configuration is executed using optical components or a holographic memory and a laser array. Although the ORGA is a multi-context programmable gate array, Fig. 1 portrays a single-context ORGA. A photograph is presented in Fig. 2. Figure 3 shows that the holographic memory was constructed using a film, on which the holographic memory pattern shown in Fig. 4 was recorded. Figure 5 shows that the holographic memory film can generate a configuration context pattern of an AND circuit. Such a two-dimensional optical configuration context pattern is programmed onto the ORGA-VLSI. The total gate count of the ORGA-VLSI is 8,704.

III. RADIATION-HARDENED AND STABILIZED POWER SUPPLY UNIT

We have developed a radiation-hardened stabilized power supply unit based on lithium-ion battery cells. The circuit diagram is shown in Fig. 8. The lithium-ion battery cell is shown in Fig. 7. The 2600-mAh capacity lithium-ion battery cell can generate 3.785 V output voltage. For this experiment, to support continuous 24 hr ORGA operation, two cells are connected in parallel. Additionally, the parallel-connection-batteries are connected in serial to produce an ap-

978-1-5386-1202-6/19 $31.00 © 2019 IEEE

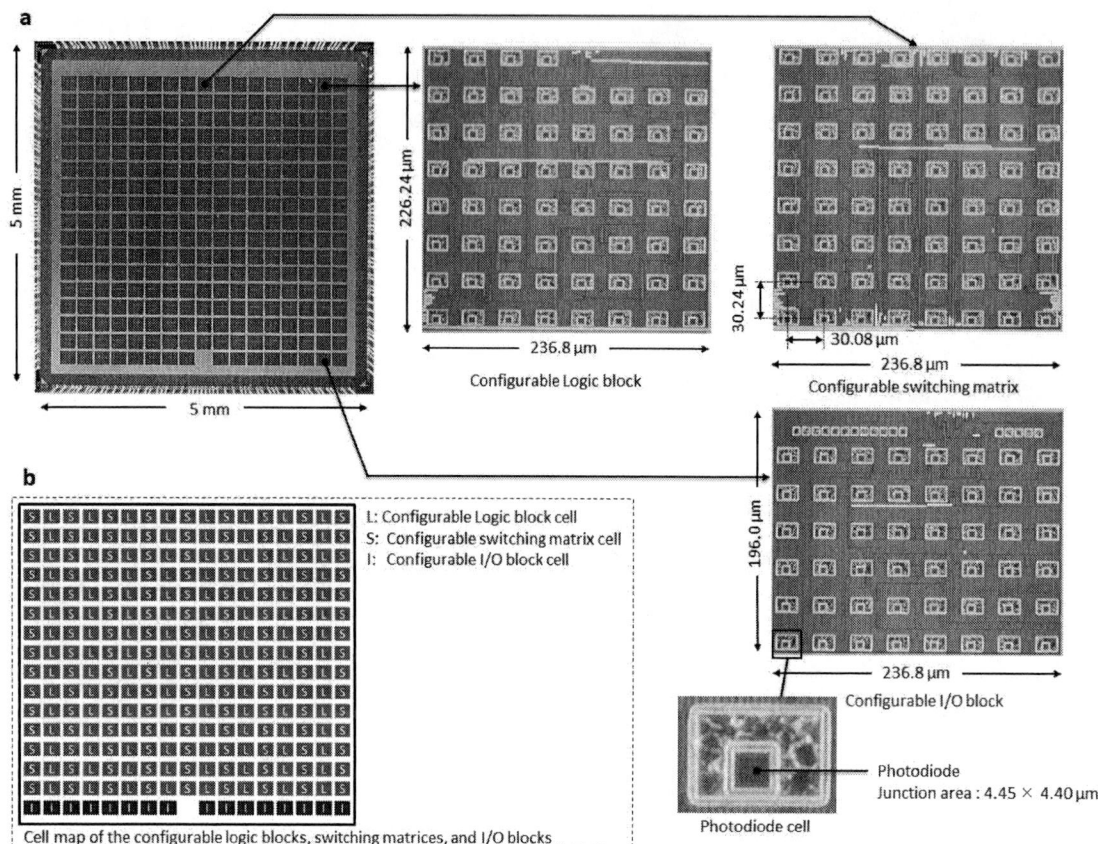

Fig. 6. Optically reconfigurable gate array VLSI chip: a, Die photograph of the 5 mm x 5 mm chip and expansion photographs of a configurable logic block (L), a configurable switching matrix (S), a configurable I/O block (i), and a photodiode cell. b, Cell map presenting locations of the configurable logic blocks, configurable switching matrices, and configurable I/O blocks.

Fig. 7. Lithium-ion battery cell.

proximately 7.5 V output voltage, which is supplied for a radiation-hardened stabilized power supply unit. The radiation-hardened stabilized power supply unit was constructed using bipolar transistors and bipolar-transistor-based comparators (LM358N) since field effect transistors are invariably vulnerable to radiation. Since the ORGA-VLSI requires 1.8 V core voltage and 3.3 I/O voltage, the radiation-hardened stabilized power supply unit is able to generate two voltages or 1.8 V and 3.3 V. A laser is also activated using 3.3 V. First, the four lithium-ion battery cells were sufficiently charged. Subsequently, the ORGA was powered continuously by the four lithium-ion battery cells and radiation-hardened stabilized power supply unit with no recharge. The ORGA and the stabilized power supply unit with the four lithium-ion battery cells were exposed to Co60 gamma radiation at a dose rate of 15.3-22 rad/s (Si). Results of this experiment demonstrated that an ORGA powered by the radiation-hardened power supply unit can function correctly for 24 hr at a dose rate of 15.3-22 rad/s (Si) using a Co60 gamma radiation source. The ORGA was able to function correctly under such a strong radiation environment.

IV. CONCLUSION

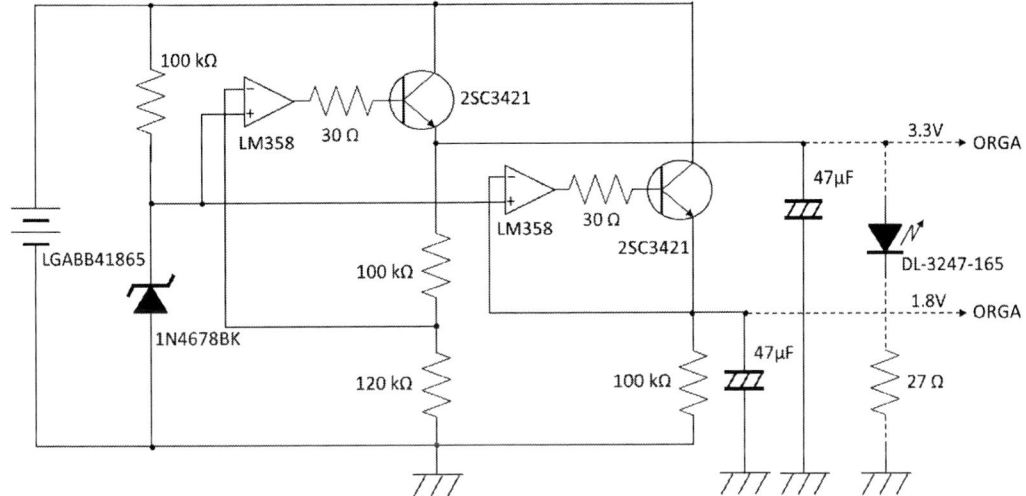

Fig. 8. Circuit diagram of the radiation-hardened and stabilized power supply unit.

Optically reconfigurable gate arrays have been under development for use as radiation-hardened devices. Although the ORGAs are a type of field programmable gate array (FPGA), by introducing an optical holographic memory technology onto currently available semiconductor technology, the radiation tolerances of ORGAs can be increased drastically. The total-ionizing-dose tolerance of the ORGA VLSI has reached over 400 Mrad, which is 400 times higher radiation tolerance than current radiation-hardened VLSIs. Moreover, we have developed a radiation-hardened power supply unit based on lithium-ion battery cells. From the experiment using a Co60 gamma radiation source described herein, we have demonstrated that an ORGA powered by the radiation-hardened power supply unit can work correctly for 24 hr at a dose rate of 15.3-22 rad/s (Si). Results clarify that the ORGA can serve in current and future decommissioning operations conducted at the Fukushima Daiichi nuclear power plant.

ACKNOWLEDGMENTS

This research was partly supported by the Initiatives for Atomic Energy Basic and Generic Strategic Research No. 283101, the Ministry of Education, Science, Sports and Culture, Grant-in-Aid for JSPS Research Fellow, No. 16J12063 and Grant-in-Aid for Scientific Research(B), No. 15H02676. The VLSI chip in this study was fabricated in the chip fabrication program of VLSI Design and Education Center (VDEC), the University of Tokyo in collaboration with Rohm Co. Ltd. and Toppan Printing Co. Ltd.

REFERENCES

[1] I. Gunawan, A. Gorod, L. Hallo, T. Nguyen, "Developing a system of systems management framework for the Fukushima Daiichi Nuclear disaster recovery," International Conference on System Science and Engineering, pp. 563-568, 2017.

[2] M. Nancekievill, A. R. Jones, M. J. Joyce, B. Lennox, S. Watson, J. Katakura, K. Okumura, S. Kamada, M. Katoh, K. Nishimura, "Development of a Radiological Characterization Submersible ROV for Use at Fukushima Daiichi," IEEE Transactions on Nuclear Science, Vol. 65, Issue 9, pp. 2565 - 2572, 2018.

[3] L. T. Clark, D. W. Patterson, C. Ramamurthy, K. E. Holbert, "An Embedded Microprocessor Radiation Hardened by Microarchitecture and Circuits IEEE Transactions on Computers, Vol. 65, Issue 2, pp. 382-395, 2016.

[4] R. Trivedi, N. M. Devashrayee, U. S. Mehta, N. M. Desai, H. Patel, "Development of Radiation Hardened by Design(RHBD) primitive gates using 0.18 um CMOS technology," International Symposium on VLSI Design and Test, pp. 1-2, 2015.

[5] T. Fujimori, M. Watanabe, "Parallel light configuration that increases the radiation tolerance of integrated circuits,"Optics Express, Vol. 25, Issue 23, pp. 28136-28145, 2017.

[6] T. Fujimori, M. Watanabe, "High-speed scrubbing demonstration using an optically reconfigurable gate array,"Optics Express, Vol. 25, Issue 7, pp. 7807-7817, 2017.

[7] Y. Ueno, M. Watanabe, "Fiber remote configuration for an optically reconfigurable gate array with four configuration contexts," Optics Communications, vol. 283, issue 23, pp. 4614-4618, 2010.

[8] T. Mabuchi, M. Watanabe, "A superimposing acceleration and optimization method of optical reconfiguration speed without any increase of laser power," Applied Optics, vol. 49, no. 22, pp. 4120-4126, 2010.

[9] M. Watanabe, T. Shiki, F. Kobayashi, "Scaling prospect of optically differential reconfigurable gate array VLSIs," Analog Integrated Circuits and Signal Processing, vol. 60, issue 1-2, pp. 137-143, 2009.

978-1-5386-1202-6/19 $31.00 © 2019 IEEE

Poster Session

978-1-5386-1202-6/19 $31.00 © 2019 IEEE

Posters

Poster Abstracts

Atomic Layer Deposition of Gallium Phosphide

Sara Kuraitis and Elton Graugnard

Micron School of Materials Science and Engineering, Boise State University, Boise, ID 83725 USA (emails: saragoltry@u.boisestate.edu and eltongraugnard@boisestate.edu).

Abstract

Atomic layer deposition (ALD) is a chemical vapor deposition technique used to grow thin film materials with sub-nanometer thickness precision. Typical binary (AB) ALD processes include the cyclic self-limiting surface reactions of chemical precursors for elements A and B, which are introduced into the deposition chamber in sequential vapor pulses separated by purges. With the proper choices of chemicals and processing conditions, these reactions saturate, leading to the self-limiting nature of ALD, which allows for film thickness to be controlled on the atomic level by varying only the number of ALD cycles. ALD of III-V semiconductors has been widely studied and ALD of gallium phosphide (GaP) is of particular interest due to its high bandgap (2.26 eV) and high refractive index (n>3) across the visible spectrum. Although preliminary results have been reported for GaP ALD using lower toxicity alternatives to phosphine, such as tris(dimethylamino)phosphine (TDMAP) and tertiarybutylphosphine (TBP), the chemical reactions and half-cycle pathways of GaP ALD with these precursors remain unknown. Toward addressing these gaps, we present in situ and ex situ process metrology to elucidate the reaction mechanisms for GaP ALD for several different chemistries, such as trimethylgallium (TMGa) and phosphine alternatives such as TDMAP or TBP as phosphorus sources. The results from these studies provide insight into the fundamental chemical reactions for GaP ALD, which has wide-ranging applications including solar cells, transistors, waveguides, photonic crystals, and other photonic devices.

Index Terms: Atomic layer deposition, Gallium Phosphide, Semiconductor Films, Materials Synthesis

Acknowledgement

This work was supported in part by the Micron School of Materials Science and Engineering.

Posters

Investigation of Molybdenum Disulfide Atomic Layer Deposition on Dielectric Surfaces

Steven Letourneau, Jake Soares, Anil U. Mane, Jeffrey W. Elam, and Elton Graugnard

S. Letourneau, A.U. Mane, and J.W. Elam are with the Applied Materials Division, Argonne National Laboratory, Argonne IL 60439 USA (emails: sletourneau@anl.gov, amane@anl.gov, jelam@anl.gov).

J. Soares and E. Graugnard are with the Micron School of Materials Science and Engineering, Boise State University, Boise, ID 83725 USA (emails: jakesoares@u.boisestate.edu and eltongraugnard@boisestate.edu).

Abstract

Widely used in the microelectronics industry, atomic layer deposition (ALD) is a thin film growth technique that allows for the conformal deposition of material through sequential, self-limiting surface reactions. Atomic-layered molybdenum disulfide (MoS_2) is a two-dimensional semiconducting transition metal dichalcogenide with a ~1.8 eV direct band gap in its single layer form. In addition to synthesis via chemical vapor deposition, ALD processes have been reported for MoS_2 using a variety of chemical precursors. However, the fundamental surface reactions involved in nucleation and growth of the MoS_2 films is not well understood. Here, we report studies of MoS_2 ALD using molybdenum hexafluoride (MoF_6) and hydrogen sulfide (H_2S). As a typical gate dielectric in field effect transistors, aluminum oxide (Al_2O_3) was used as the growth surface. In situ and ex situ characterization was performed to gain insight into the initial surface reactions. Understanding the chemistries and reactions between the target surface oxide and precursor reactants, during the first few ALD cycles, allows for a better evaluation of MoS_2 nucleation during the deposition and its ability to form the desired layered structure.

Index Terms— Atomic layer deposition, Molybdenum Sulfide, Semiconductor Films, Materials Synthesis

Acknowledgement

This work was supported in part by the Micron School of Materials Science and Engineering, NSF grant No. 1751268. S.L. acknowledges support from the U.S. Department of Energy, Office of Science, Office of Workforce Development for Teachers and Scientists, Office of Science Graduate Student Research (SCGSR) program. The SCGSR program is administered by the Oak Ridge Institute for Science and Education for the DOE under Contract DE-SC0014664.

Posters

Dependency of RGB and metrology data correlation on wafer alignment

Swetha Barkam, James Cultra, Zhenyu Bo, Sri Sai Vegunta

Micron Technology Inc., Boise, Idaho 83716 USA (email: sbarkam@micron.com)

Abstract

Chemical Mechanical planarization is crucial for semiconductor device fabrication in eliminating excess film deposition. It is vital to have a robust inline detection technique for underpolish/over-polish (OP/UP) issues that cause subsequent downstream issues leading to yield fallout. Additionally, the robust inline control is more significant for high volume manufacturing for quality control. Currently, metrology data is collected inline to detect UP/OP, however not all dies and features on the wafer get metrology data. In general, optical images are collected post CMP process to detect gross UP/OP, defectivity, etc. In this study, we use an integrated imaging technology to collect optical images of the wafers and their subsequent RGB intensity data. The RGB intensity values will be correlated to the metrology data to obtain the film thickness for each pixel of the image, including the regions/dies/structures that did not receive metro measurement. The RGB intensity data depends on multiple factors such as thickness of the film, incident angle of light source, refractive index of the film, critical dimension of pattern features etc. Additionally, the imaging system rotates the wafer to a specific position prior to obtaining the image, which consumes time causing capacity issues for fab operations. In this study, we shall explore how RGB intensity data is affected by the wafer rotation angle while capturing the image. This in turn effects the correlation factor of RGB intensity and metrology data with change in wafer rotation angle.

Index Terms: CMP, Virtual Metrology, Within-Wafer Non-Uniformity, Under-polish and Over-polish detection capability.

Posters

Scanning Probe Microscopy for Nanoscale Characterization of Electrical and Magnetic Properties

Olivia Maryon

Boise State University, Boise, ID 83725 USA (emails: oliviamaryon@u.boisestate.edu)

Atomic force microscopy (AFM) is a nanoscale scanning probe microscopy (SPM) characterization technique useful for obtaining topographical maps of surfaces and their associated nanomechanical properties. Complementary SPM modes such as Kelvin probe force microscopy (KPFM) and magnetic force microscopy (MFM) can simultaneously elucidate the electrical and magnetic properties of materials with nanoscale resolution, thereby expanding AFM's utility. KPFM measures the Volta potential difference between a conductive AFM probe and the sample surface, which can be related back to the work function of the material and correlated with co-localized elemental mapping via energy dispersive spectroscopy (EDS). This can be useful for understanding and predicting initiation and propagation of galvanic corrosion in metal alloys. MFM employs a magnetized AFM probe tip to detect magnetic interactions between the sample and the tip, thereby mapping out the magnetic structure of the sample surface. Here we present KPFM studies of case-hardened stainless steels engineered for bearing applications in high performance jet engines destined for operation in corrosive marine environments. MFM studies of Ni-Mn-Ga, a magnetic shape memory alloy, connect experimental data with computational modeling to understand the growth of twins in response to bending. Together, these studies highlight the widespread applicability of AFM, KPFM, MFM, and other SPM techniques for illuminating nanoscale structure-property relationships in material systems.

IEEE WMED 2019 Author Index

A

Abugri, Joseph B. 10

B

Barkam, Swetha 40
Bo, Zhenyu 40
Budhathoki, Sujan 10
Butte, Sujata 18

C

Carnevale, Gianpietro 1
Cultra, James 40
Chen, Y. 5
Cho, Hyejin 1
Clark, Billy D. 10
Crowell, Ethan 27

D

Di, Mengfu xxv

E

Elam, Jeffrey W. 39
Estrada, David xxi

F

Fayrushin, Albert 1
Fujimori, Takumi 32
Fujisaki, Shinya 32

G

Graugnard, Elton 39
Groothuis, Steve xx
Gupta, Subhadra 10

H

Hauser, Adam 10
Houle, Jennifer 27

K

Kang, Sung-Mo Steve 23
Kotti, Radhakrishna 15
Koushan, Foroozan S. 23
Kulkarni, Jaydeep xxii
Kuraitis, Sara 39
Kuzyk, Mark G. 27

L

Lawson, M. 5
Letourneau, Steven 39
Li, L. 5
Li, Cheng xxv
Li, S. 5
Liu, Haitao 1

M

Mane, Anil U. 39
Mao, Duo 1
Maryon, Olivia 41
Mauri, Aurelio 1
Mossman, Sean 27

P

Pan, Zijin xxv
Pop, Eric xix

S

Soares, Jake 39
Strukov, Dmitri B. xv
Sullivan, Dennis 27

V

Vakanski, Aleksandar 18
Vegunta, Sri Sai 40
Visscher, Pieter B. 10

W

Wang, Albert...xiv,xxv

Wang, Han xxv

Watanabe, Minoru 32

Williamson, I. 5

Wuttig, M. xvii

X

Xian, Min 18

Z

Zhang, Feilong xxv

Zhao, Kejie... xviii

2019 Sponsors

Executive Sponsors

BOISE SECTION

BOISE CHAPTER

Platinum Sponsors

Gold Sponsors

Educational Sponsors

978-1-5386-1202-6/19 $31.00 © 2019 IEEE

IEEE
445 Hoes Lane
Piscataway, NJ 08854-4141

ISBN 978-1-5386-1202-6